THE NATURE
OF SOLIDS

THE NATURE
OF SOLIDS

ALAN HOLDEN

DOVER PUBLICATIONS, INC.

New York

MAY 0 3 2018

This Dover edition, first published in 1992, is an unabridged, unaltered republication of the work first published by the Columbia University Press, New York, 1965. This edition is published by special arrangement with Columbia University Press, 562 West 113 Street, New York, NY 10025.

Library of Congress Cataloging-in-Publication Data

Holden, Alan.
    The nature of solids / Alan Holden.
        p.   cm.
    Reprint. Originally published: New York : Columbia University Press, 1965.
    Includes index.
    ISBN-13: 978-0-486-27077-7
    ISBN-10: 0-486-27077-7
    1. Solids.   I. Title.
QC176.H6   1992
530.4′1—dc20
                                                              91-38931
                                                              CIP

Manufactured in the United States by RR Donnelley
27077708   2016
www.doverpublications.com

FOREWORD

THIS BOOK provides to anyone who has studied physics or chemistry in a secondary school a nonmathematical account of how some of the behavior of solid matter is understood at present. Such a reader will find his prior knowledge rehearsed in early chapters and brought to bear on a few of the central questions raised by any penetrating study of solids. In later chapters he will learn why some of those questions cannot be answered in the terms usually taught in secondary schools, and he will meet the wave-mechanical pictures now used to answer them.

Necessarily this book gives a distorted view of the scientific adventure. Science was made by men, and men will continue to make it. Some of superior intellect and many more of ordinary ability have constructed the edifice patiently—making mistakes and correcting them—adding bricks, some marked with their names and others anonymous. The history of their efforts must be read elsewhere.

So too must any adequate description of the evidence for the pictures that this book presents. Like all scientific theories, they derive their final support from facts, often observable only in experiments ingeniously contrived to confirm or deny the theories. Only a few hints, however, of the interaction between theory and experiment appear in these pages. Instead they are devoted to explaining the theories—to picturing the models—that provide the best means known today for unifying our knowledge of solids and connecting it with broader fields of science. The study of that knowledge is commonly divided between the specialties of physicists, chemists, and crystallographers. Bringing together here their separate ways of speech may help to make clear the unity to which the scientific effort aspires.

ALAN HOLDEN

Murray Hill, New Jersey
January, 1965

CONTENTS

THE NATURE OF SOLIDS

I believe that natural science can be made accessible, and that the specialized apparatus of formulae and experimental technique, unavoidable though they be for the expert, can be dispensed with when the reader's aim is, by a receptive survey of the field, to obtain final insight into the meaning and results of scientific work.

HANS REICHENBACH, *Atom and Cosmos*

I. THEORIES

We make to ourselves pictures of facts.
LUDWIG WITTGENSTEIN,
Tractatus Logico-philosophicus

FORTUNATELY we have no trouble walking through a gas: air yields easily. A liquid looks more forbidding, and you will find it so if you jump into a pool and land flat on your stomach. But many a baby, confident that he can walk on water, is dismayed to find that it lets him down.

A solid gives a wholly different account of itself. It is rigid, and in order to get through it, you must hit it hard enough to break it. Once broken, a solid, unlike a liquid or a gas, stays broken. But to break it enough to walk among the pieces, as you would walk in a liquid or a gas, would take a great deal of pounding.

Such facts impressed Sir Isaac Newton sufficiently to draw from him a memorable demand:

The Parts of all homogeneal hard Bodies which fully touch one another, stick together very strongly. And for explaining how this may be, some have invented hooked Atoms, which is begging the Question . . . . I had rather infer from their Cohesion, that their Particles attract one another by some Force, which in immediate Contact is exceeding strong, at small Distances performes the chymical Operations above-mention'd and reaches not far from the Particles with any sensible Effect . . . . There are therefore Agents in Nature able to make the Particles of Bodies stick together by very strong Attractions. And it is the Business of experimental Philosophy to find them out.*

Optics (1704).

Before experimental philosophy could undertake Newton's "business," much other work had to be done. Heat and light, just as familiar as solids, were just as mysterious. Few men recognized the existence of electricity other than lightning. Two centuries elapsed before the emerging physical and chemical scheme for interpreting Nature could begin to offer a penetrating response to Newton's challenge.

By that time many other properties of solids nagged the curious as much as their cohesion did. Why are some solids transparent and others opaque, some electrical conductors and others insulators? Why will some absorb much heat and others little? To describe these properties—to measure them and affix numbers to them—only sharpened the appetite to explain them.

There emerged two distinguishable ways of setting out to explain the behavior of solids: the *macroscopic* way and the *atomistic* way. Today these two ways supplement each other, and it is worth while to look at how they both work.

The Macroscopic Approach

Before much was known about atoms and their behavior, only the macroscopic approach was available. This way of thinking, sometimes called the *phenomenological* way, does not inquire into the ultimate construction of a solid. It reaches its conclusions without needing to know that construction.

The thinking shows, for example, that any solid bit of matter must necessarily contract, much or little, when pressure is applied to it. Steel, wood, glass—all must obey that rule. You may say, "Of course. What else would you expect?" But before you brand the rule trivial, examine another example. Solids usually contract when they are cooled. Must all solids contract when cooled, as they must when pressed? No, some expand.

When no phenomenological rule asserts that a usual sort of behavior is a necessary sort of behavior, we can expect to find exceptions to the rule. Nature is so diverse that she can provide exceptions to any but the rigorous rule. But the macroscopic approach, from which a solid appears to be a structureless piece of matter, can do little to help in finding the exceptions and nothing to explain them.

The little that it can do, however, cannot be neglected. If a solid exhibits one sort of exceptional behavior, then that behavior may imply that the solid is necessarily exceptional in some other way as well. In other words, there may again be a phenomenological rule asserting that *if* so-and-so is true, *then* such-and-such must be the case.

For example, most liquids contract when they freeze. But everyone whose plumbing has burst in cold weather knows that water is exceptional. Ice takes up more space than the water from which it was formed and floats on water for that reason. Conversely, although it is less obvious, pressure makes ice melt at a temperature lower than its usual melting point. These two kinds of behavior have a necessary connection; either implies the other.

Thus, the familiar observation that ice floats on water enables us to predict the unexpected fact that ice under pressure will melt at a lower temperature. The ice skater glides on a thin film of water produced by the weight of his body, and the water freezes again when he has passed.

This behavior exemplifies a particular class of the properties of matter that can be approached with great success macroscopically. Those properties are called *reversible*. To see how reversible properties are distinguished from other kinds, look at how some familiar mechanical properties of solids might be classified.

Reversibility, Transport, and Catastrophe

When a solid object is pulled in opposite directions on opposite ends, when it is squeezed, twisted, or bent, a *stress* is applied to it. The solid responds to the stress by acquiring a *strain*, a change in its size or shape. A uniform pressure is a special kind of stress, and the resulting decrease of volume is a special kind of strain.

If the stress is sufficiently small, removing the stress will remove the strain — the solid will spring back to its original shape. In short, this *elastic* behavior of solids is reversible, and the elasticity of solids is a reversible property.

A larger stress may exceed the *elastic limit* of the solid so that when the stress is removed, the solid springs only part of the way back. The permanent change of shape is a consequence of the *plastic flow* of the material. A still larger stress will make a *cata-*

strophic change in the solid — the piece will break apart. Flow and fracture are *irreversible* properties — the effects remain after their causes have disappeared.

It is interesting to classify some of the electrical properties of solids in an analogous way. If two pieces of metal are placed on opposite sides of a thin sheet of mica and then connected by wires with the terminals of a battery, the mica will not conduct an electric current continuously. But as soon as the pieces of metal are connected with the battery, a pulse of current will *polarize* the mica electrically — or charge it, to use the popular word. If the wires are then disconnected from the battery and touched together, there will be a pulse of current through them in the opposite direction, discharging the mica.

Discussing this phenomenon a hundred years ago, James Clerk Maxwell wrote,

Here, then, we perceive another effect of electromotive force, namely, electric displacement, which according to our theory is a kind of elastic yielding to the action of the force, similar to that which takes place in structures and machines owing to the want of perfect rigidity of the connexions.*

Such *dielectric* polarization occurs also in all electrical insulators; it is a reversible electrical property analogous to their mechanical elasticity.

A stronger battery may drive a continuous trickle of electric current through an insulator, especially if the insulator is hot. In other words, a high enough voltage slowly pushes electric charges through the solid, somewhat as a high enough stress slowly transports some of the material into a new position. Finally, a very high voltage will drive a spark through the solid, leaving a little trail of catastrophe behind it.

Some thermal properties of solids can be classified in similar terms. When a hot object is put in contact with a cold object, heat flows from the hot object into the cold object until their temperatures are the same. If the newly heated object is then put in contact with another cold object, the heat will flow out of it again. Each

Philosophical Transactions of the Royal Society (London), Series A, CLV (1865), 459.

object has a capacity for absorbing heat, and under suitable circumstances it will return the heat that it absorbs. You can think of a difference of temperature as somewhat like a force that drives heat. Heat capacity thus stands out as a reversible property and heat conduction as a *transport* property.

Heated to a high enough temperature, a solid object may melt, absorbing quite a lot of heat as it does so. If the molten material is cooled, it will solidify again and return the heat. From that point of view, fusion is a reversible process. But from another point of view it is catastrophic, for the solid stops being solid when it melts and must be made anew. This second point of view is shown in Table 1, which summarizes the preceding classification.

TABLE 1. A Suggested Classification of Some Properties of Solids

	Reversible	Transport	Catastrophic
Mechanical	Elastic distortion	Plastic distortion	Rupture
Thermal	Heat capacity	Heat conduction	Melting
Electrical	Dielectric polarization	Electrical conduction	Dielectric breakdown

Both the macroscopic and atomistic approaches treat the reversible properties of solids more successfully than the irreversible properties. The theory of elasticity is a good example of the macroscopic study of a single reversible property. In that theory, mathematical methods are employed to deduce many consequences of two simple observations.

The first, sometimes called *Hooke's law*, is the fact that the strain in a solid is proportional to the stress producing it, so long as the solid is not stressed beyond its elastic limit. The second observation is the fact that the mechanical work done on a solid, by stressing it below its elastic limit, can be completely recaptured as mechanical work when the stress is removed. The deduced behavior of solid objects under static loads and in vibration is a cornerstone of much modern mechanical engineering.

Couplings and Their Converses

But applying stresses is not the only way of producing strains. The mechanical, thermal, and electrical properties of a solid do not live in isolation from one another. The expansion of a solid when it is heated is an instance of a thermal influence producing a mechanical effect: a change of temperature produces a strain. There are similar couplings between all these properties, and they are especially interesting when the properties coupled are reversible.

About two hundred years ago Franz Aepinus noticed that when the gem stone tourmaline is heated, it becomes electrified. Later work showed that a change of temperature polarizes tourmaline electrically in much the same way that an electric battery polarizes mica. In tourmaline and in many other crystals — even crystals of ordinary sugar — a thermal influence produces an electrical effect called the *pyroelectric effect*.

In 1880 the brothers Pierre and Jacques Curie discovered that many crystals exhibit another curious coupling between their reversible properties. Quartz and Rochelle salt crystals are instances of materials that can be electrically polarized by a mechanical stress. This *piezoelectric effect* has had wide application — in phonographs, for example — for translating mechanical motions into electrical impulses. Thus, the circle of couplings (Fig. 1) between

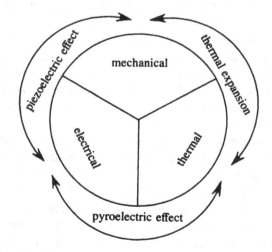

FIG. 1 — Couplings between the reversible properties of a solid.

the reversible mechanical, thermal, and electrical properties of solids has been closed by instances of the three possible types.

You may feel that the circle is not quite closed until you have found solids that behave conversely—instances in which an electrical influence produces a mechanical or a thermal effect. But here the macroscopic theory provides the answer that you have found such instances already. This is typical of the answers that the theory can give regarding couplings between reversible properties.

If, for example, a solid exhibits the piezoelectric effect—in which a mechanical influence produces an electrical result reversibly—then that same solid will exhibit the converse effect—that is, an electrical influence will produce a mechanical result. A voltage applied to the crystal will change its shape slightly; when the voltage is removed, the crystal will return to its original shape. Indeed, if you measure the magnitude of either of these effects, you need not measure the other. You can calculate either from its converse by using the macroscopic theory, without recourse to an atomistic approach.

The Conservatism of Matter

Even without calculation we can get some useful qualitative ideas of how the macroscopic theory connects the reversible coupling properties of matter. One connection can be made by a powerful generalization, often called the *principle of mobile equilibrium,* stated by Henri Le Châtelier in 1884 and sometimes called the *Le Châtelier principle.*

This principle says that matter resists change: when it is forced to change, it opposes the force with all the means at its disposal. You can see most clearly how to use the principle by recalling the behavior of a gas under pressure.

When a gas is compressed, it is forced to occupy a smaller volume. It must comply, but it complies as reluctantly as it can. A rise in its temperature would tend to make its volume increase and so help oppose the force. Therefore, as anyone who operates a tire pump will notice, compressing a gas raises its temperature.

Similarly, the Le Châtelier principle connects the properties of ice discussed above. Imagine a mixture of ice and water just at the freezing temperature, so that none of the ice is melting and none of the water is freezing. Pressure exerted on the mixture forces it to occupy a smaller volume. Since water occupies a smaller volume than ice, some of the ice is forced to melt into water, so that the volume is reduced. At the same time the temperature will go down, in an effort to keep the ice frozen and prevent the decrease in volume. Here is a case that behaves opposite to the gas in a tire pump; pressure reduces the temperature instead of raising it.

You may be interested in performing two experiments on two analogous properties of an ordinary rubber band. For the first experiment, hold the ends of the band in both hands and stretch it to four or five times its normal length. Keep it stretched for a minute or so—long enough to insure that its temperature is the same as that of the room—then let it retract to its normal length. Now quickly touch the tip of your tongue, a sensitive thermometer, to the rubber band. Is it hotter or colder than you would expect? If you are in doubt, compare it with a similar rubber band that you have not stretched.

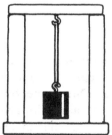

FIG. 2—A rubber band stretched by a weight.

For the second experiment, arrange a weight so that it hangs freely from the rubber band (Fig. 2). Measure the length to which the weight stretches the band, and then put the whole assembly in the refrigerator. After giving the band a few minutes to cool, measure its length again while it is still cool and compare the new length with the previous length. By the Le Châtelier principle, you can connect these observations of length with the observations of temperature in the first experiment.

The Atomistic Approach

Note, however, that reasoning of this kind cannot predict the magnitudes of the effects that you observe. It can only connect the two facts that the rubber band is cooler when you have finished your first experiment and that it becomes longer in the refrigerator in your second experiment. And the results of the theory of elasticity can contain only algebraic symbols for the constants of proportionality between stress and strain. The only way to substitute a number for a symbol is to measure the ratio of the strain to the stress in a sample of a material in question. Taken alone, the theory lacks names and numbers.

An ideal of the second approach to solids—the *microscopic*, or *atomistic*, approach—is to add those names and numbers—to penetrate more deeply into the behavior of solids and to deduce their observed properties from a knowledge of the way they are constructed by their constituent atoms and from a knowledge of the physical behavior of those atoms. The history of this approach extends back more than a hundred years. Indeed, an atomic constitution of matter had been suspected for two millenia. But the accomplishments of the atomistic approach were meager until, in this century, the discovery of X-ray diffraction by crystals and the invention of wave mechanics opened avenues for systematic progress.

The idea that all forms of matter are made of atoms provides at once a picture of the differences among the three states of matter—gaseous, liquid, and solid. In gases the atoms (or perhaps the molecules, each consisting of a few atoms tightly tied together) are flying about as tiny, self-sufficient units. They move quite independently and influence one another very little, except during the instants when two collide and one picks up speed at the expense of the other. If the gas were not confined in a container, the molecules would fly off in all directions.

When the temperature of the gas is reduced, the average speeds of the molecules are reduced also, and the molecules have a chance to respond to the attractive forces they exert on one another. At a low enough temperature, those forces will bring them together to form a liquid. The molecules are packed quite tightly in a liquid, but they are still moving. In particular, they can move past one another and permit the liquid to flow.

At an even lower temperature, the liquid solidifies. In the solid the molecules are packed only very little more tightly than in the liquid (indeed, in ice they are packed *less* tightly), but they can no longer move past one another easily. Their motion is a vibration about fixed positions from which they do not stray far. Even at a temperature of absolute zero, were it attainable, the molecules would continue to vibrate a little, with the so-called *zero-point vibrations*.

In the picture of solids as built of molecules tightly packed together in mutual attraction and subject to constant thermal agitation, it is clear why a quantitative atomistic theory of solids is difficult to perfect. When constructed from first principles, mathematical equations describing the behavior of a collection of particles as unimaginably numerous as the atoms in a solid become too complicated to solve and too opaque to give insight. Characteristically, the physics of solids retreats from the ideal of deducing the exact quantitative behavior of solids from first principles and sets itself a more modest goal.

The Atomistic Goal

That goal is insight rather than exact calculation. In pursuing it, the physicist first separates out a particular phenomenon for study and makes an educated guess at what atomic behavior is largely responsible. He then sets up an imaginary model which embodies that behavior and neglects all the confusing details he thinks may be unimportant. He calculates how the model will behave and compares his answer with the results of experiments on actual solids. If his result is in rough agreement with experiment, he accepts his model as a correct explanation of the phenomenon, that is, a simplified approximation of it.

Clearly, several factors must conspire to make this effort successful. In his educated guess, the physicist tentatively builds an additional insight on the past insights of others, with which he has had to make himself acquainted. His model must be consistent with other successful models used in discussing other phenomena, for although the phenomena may be different, the solid in which they are occurring is the same. In constructing the model, the physicist must be adroit enough to make it mathematically manageable, so that he can get a numerical answer for the crucial comparison with experiment. And he must use good physical sense in deciding what degree of agreement with experiment is *good* agreement; in some cases good agreement may mean within 10 percent, in others within a factor of 10.

Throughout most of this book you will see the atomistic method at work. And you will notice that the models are built with no more detail than needed to describe the behavior they were invented to explain. This habitual economy obeys a deep-lying aesthetic principle of all the sciences, often called "Occam's razor": Let not hypotheses be multiplied beyond those necessary to explain the facts. A physicist will try to remove from a model those details which are irrelevant to his purpose. And he will try to avoid using an atomistic model for a purpose that could have been served as well without the assumption that matter is composed of atoms, that is, by a macroscopic argument.

The earliest and simplest atomistic models of a solid—those devised for explaining heat capacity—are discussed in the next two chapters. Their simplicity and success have combined to put them in an especially important position among solid models.

II. HEAT

*It is odd to think that there is a word for something which
strictly speaking does not exist, namely, rest.*
MAX BORN, *The Restless Universe*

DURING the nineteenth century, physicists succeeded in their quest for precise ideas about the two major thermal quantities—heat and temperature. In earlier days it had seemed natural to suppose that the heat which flowed from a hot to a cold body was some unique substance. But about the year 1800 many experiments, especially those of Count Rumford, showed that the heat in hot bodies is associated with a mode of *motion*—a disorganized random motion of their constituent molecules.

Thus, the heating of a cold body is now visualized as an excitation of the motion of its molecules by the impacts of the molecules in the hot body. Both the heat that flows and the temperature differences that urge it to flow can be related to those molecular motions.

The Energy of Heat

Even before 1800 many measurements had been made of the amount of heat required to change the temperatures of various substances by various amounts, and *heat capacity*—the amount of heat necessary to raise the temperature of a given mass of the substance by one degree—had been tabulated. In trying to bring order to these data, Pierre Dulong and Alexis Petit made an important observation shortly after Count Rumford's death. They noticed that most of the chemical elements in solid form absorbed nearly the same amount of heat, regardless of the species of the element, if the weights of solid taken for comparison were proportional to the chemist's atomic weights of the elements composing them.

The *atomic weights* of the chemical elements are numbers that describe the relative proportions in which the various atomic species combine with one another to form chemical compounds. The atomic weights assigned to carbon and oxygen, for example, are approximately 12 and 16. Accordingly, 12 grams of carbon will combine with 16 grams of oxygen to form 28 grams of carbon monoxide or with twice as much oxygen to form 44 grams of carbon dioxide.

Chemical compounds always contain simple proportions of their constituent elements when these atomic weights are chosen as units of weight for the elements. This fact strongly supports the idea that matter is constructed of atoms. According to the atomic hypothesis, each chemical compound is made of a myriad of identical tiny molecules. Each molecule is made in turn of relatively few atoms, and the actual weights of the atoms are proportional to the chemist's atomic weights.

Hence, the number of carbon atoms in 12 grams of carbon is the same as the number of oxygen atoms in 16 grams of oxygen. And that number is the same as the number of carbon dioxide molecules in 44 grams of carbon dioxide. The number of atoms in a *gram-atomic weight* of an element, or the number of molecules in a *gram-molecular weight* of a chemical compound, is a quite fundamental number in physical and chemical calculations. It is called *Avogadro's number* and is approximately 6.02×10^{23}.*

If you accept the atomic hypothesis, you can see what Dulong and Petit were really pointing out: the amount of heat absorbed by a solid element depends primarily on how many atoms are present and very little on what kind of atoms they are. In other words, the heat required per atom to raise the temperature of an elementary solid is much the same for most of the elements. About 6 calories

---

*Convenience in writing and thinking about numbers that are very large and very small is promoted by notation that expresses those numbers in powers of 10. Thus, 10^{17} denotes the large number that would otherwise be written as 1 followed by 17 zeros, and 10^{-17} denotes the tiny fraction that would otherwise be written as 1 divided by that same number. In consequence, 2×10^3 denotes two thousand, and $\frac{1}{2} \times 10^{-3}$ or 5×10^{-4}, denotes one two-thousandth.

of heat will raise by one degree centigrade the temperature of a gram-atomic weight of almost any of the elements in solid form. Exceptional cases, in which those atomic heats of the solids are much less than 6 calories, are found among the elements of low atomic weight that form solids with high melting points, such as boron and carbon.

At first thought, Dulong's and Petit's observation might tempt you to conclude that the heat required to raise the temperature of matter would be the same for all sorts and conditions of matter, if you reckon heat per atom. But that is not so. An element that requires 6 calories per gram-atomic weight to become one degree warmer when the atoms form a solid requires only 3 calories when they form a gas. The heat capacities depend less on the kind of atoms than on their state of aggregation.

The first half of the nineteenth century saw a steadily clearer recognition that the heat absorbed by matter is precisely equivalent to the *energy* associated with the motion of the atoms in the matter. About the middle of the century James Joule established the exact amount of energy (described in the Appendix) that is equivalent to one calorie, often called the *mechanical equivalent of heat*. It was reasonable to expect that some purely mechanical explanation could be found for the Dulong-Petit law and for the difference between the heat capacities of solids and gases.

The explanation came with the development of statistical mechanics, especially in the minds of James Maxwell, Ludwig Boltzmann, and Willard Gibbs. They found ways to apply statistical concepts to deduce the expected behavior of an assembly of a large number of units when each of those units obeys well-known laws of mechanics. By the method that they developed, the average behavior of the units can often be found in spite of—indeed because of—the chaotic behavior of the individual units.

Independent Atoms

Before looking at how the method of statistical mechanics is applied to a solid, you will find it helpful to see how the method is applied to a gas. Notice first that the gas is pictured in a simplified model. Each atom is taken to be an independent *mass-point*; that is, a particle having mass but no size and feeling no attraction or repulsion from the other atoms, except at rare instants of collision.

This picture drastically simplifies the situation, of course. In other models of atoms, for example, each is a complicated structure made up of its nucleus and the electrons circulating around that nucleus. And in other situations the atoms behave as if they feel attractions and repulsions from one another. But those complications seem less and less important as you imagine the gas to become more and more dilute. In other words, you expect the behavior of the model to reproduce the actual behavior of the gas more closely when the gas is allowed to expand and occupy a great deal of space.

But the collisions of the atoms continue to appear, though somewhat surreptitiously, in the model. Those collisions enable the mass-points that represent the atoms to exchange energy with one another. The two atoms engaging in a collision may have different velocities and thus different *kinetic energies* (energies of motion) when they approach each other. In the collision, one may contribute some of its kinetic energy to the other, and thus both will depart from the collision with new velocities.

By the method of statistical mechanics, the average behavior of an atom can be calculated when all the atoms are exchanging energy in this way. The behavior of the whole gas is then the same as that of an average atom multiplied by the number of atoms in the gas. Part of the calculated behavior of the model turns out to be exactly that which is described by the *perfect gas law,* a combination of the laws discovered much earlier in the experiments on actual gases made by Robert Boyle, Jacques Charles, and Joseph Gay-Lussac. That law relates the pressure p, the volume v, and the absolute temperature T of a gram-atomic weight of a gas by the simple expression $pv = RT$. The constant R in this expression is the *gas constant* and is the same for all gases of whatever element (1.985 calories per degree).

But another part of the calculated behavior is more important in the present context, because it relates directly to heat capacities. The total kinetic energy of the atoms in the model of a gas made of independent mass-points is $3RT/2$ per gram-atomic weight. Thus, in order to raise the temperature of the gas by one degree, the amount of heat required is $3R/2$, and that is its heat capacity.

From the value of R just given, the heat capacity is clearly about
3 calories, which is also the value measured experimentally and just
half the value for the solids noticed by Dulong and Petit.

Atoms and Oscillators

The problem is now to set up a reasonable mechanical model for
an atomic solid that will have a heat capacity twice as great as that
for the model of an atomic gas. In the solid, the mass-points can no
longer be pictured as independent. They are packed closely together and influence one another so strongly that they cannot move
past one another. Their motion is confined to vibration about fixed
positions.

The words "vibration about fixed positions" give a clue to the
way to set up a reasonable model. Many familiar things vibrate
about fixed positions. A pendulum bob swings back and forth
through a fixed average position, its position at the bottom of its
swing. You can set a massive object suspended from a coil spring
into oscillation about its fixed rest position by pulling it and then
letting it go. Physicists often idealize these and many other cases of
vibration into instances of a very useful fiction, the *harmonic
oscillator*.

You can visualize the harmonic oscillator as a mass-point at one
end of a weightless spring, the other end of which is held fast (Fig.
1). When the mass-point is displaced from its rest position, a force
from the spring urges it back toward that position. The force is
greater the farther the mass-point is displaced, vanishes when it is
in the rest position, and reverses direction when it is on the other
side of that position. Indeed, that restoring force is *directly proportional* to the displacement of the mass-point from its rest position, a fact that gives the harmonic oscillator especially simple
properties in contrast to *anharmonic* oscillators, in which the
relation between displacement and restoring force is more complicated.

It takes energy to make the oscillator oscillate, and clearly that
energy is greater the larger the excursions that the oscillator
makes. But the energy is obviously not all in the form of kinetic
energy, at least not all the time, unlike the energy of the freely
moving mass-point in the model of a gas. As the oscillator rushes

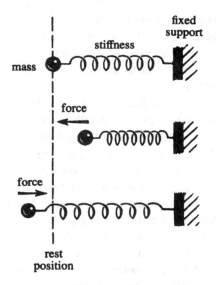

FIG. 1 — The harmonic oscillator

through its rest position — as the pendulum bob coasts through the lowest point of its swing — its energy is all kinetic. But when it reaches the farthest point of its swing, it is instantaneously at rest, and an object at rest has no kinetic energy at all.

Potential Energy and Kinetic Energy

If you think only of the kinetic energy, you might conclude that the total energy of the oscillator varies from zero when it is at the end points of its swing to some maximum value when it is in the middle of its swing. But you will be suspicious of this answer if you have ever heard of the principle of the conservation of energy. In fact, you will quickly see how the oscillator keeps the same total energy all the time when you recall an often used definition of energy as the capacity for doing work.

When the oscillator is moving through the midpoint of its swing, it can do work by hitting something else. But when it is at an extreme of its swing, it can still do work. If the oscillator is a pendulum, the pendulum bob is higher at the ends of the swing than at the midpoint, and the bob can do work by falling to the lower position and pulling something as it falls. If the oscillator is a mass attached to a spring, the spring is extended or compressed at the ends of the swing and can do work by pulling or pushing something until it returns to normal. When you start oscillations by pulling the oscillator to its extreme position, you do work on it and give it the capacity to return that work.

The capacity of the oscillator to do work by virtue of its position is its *potential energy,* and its capacity to do work by virtue of its motion is its *kinetic energy.* You can think of the oscillator as doing work on itself by giving itself an increasing velocity as it moves toward the midpoint of its swing, and thus conserving its energy as kinetic energy. After it has passed the midpoint of its swing, you can think of the oscillator as hitting itself outward again, and thus conserving its energy as potential energy.

Thus, the total energy of the oscillator is conserved but keeps changing its form (Fig. 2) from kinetic energy to potential energy. It turns out that the total energy of the oscillator, averaged over a full swing, is just evenly divided between potential and kinetic energy.

A simple model for a solid now suggests itself. Consider the solid as made of independent harmonic oscillators that can exchange energy with one another, much as the gas model was made out of independent freely moving mass-points that can collide with one other. Strong interactions between each atom and its neighbors keep that atom in place in the solid and are represented in the

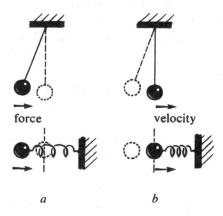

FIG. 2—(*a*) Potential energy and (*b*) kinetic energy of an oscillator.

model (Fig. 3) by the restoring force of the harmonic oscillator. When we add energy to the model by heating it, half goes into increasing the average kinetic energy of the oscillators, and the other half goes into increasing their average potential energy.

If only the kinetic energy were felt as an increase in temperature, this model would explain the fact that the heat capacity of a solid is twice as great as that of an atomic gas. A given amount of heat energy supplied to a perfect gas would all be used to increase the kinetic energy of its atoms because they have no potential energy. But if the same amount of heat energy were supplied to a solid, only half of it would appear in the kinetic energy of the oscillators, while the other half would disappear into potential energy.

Temperature

A little thought about the meaning of temperature will support this argument. First recall that the numbers ordinarily used to express temperatures are wholly arbitrary. Noticing that boiling water has a higher temperature than freezing water, you can call the temperature of freezing water 0 and the temperature of boiling water 100 degrees and thus set up the centigrade scale of temperatures.

But the only physical fact used in setting up that scale was the fact that heat will flow from boiling water to freezing water — that boiling water has the "higher" temperature. In particular, when two things have the same temperature, no heat passes from one to the other.

Now imagine a harmonic oscillator model of a solid object im-

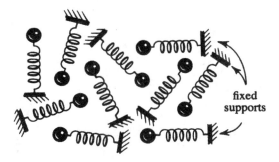

FIG. 3 — An approximate model of a solid.

mersed in a gas made of the same number of atoms. If the gas is hotter than the solid, the atoms in the gas will contribute heat to the atoms in the solid by hitting them harder than they hit back. The solid absorbs the energy of that heat by increasing the amplitude of motion of its harmonic oscillators. Since half of that energy goes into increasing the average potential energy of the oscillators and since the oscillators can hit back only with their kinetic energy, the gas will cool twice as fast as the solid warms. Thus, it takes twice as much heat to raise the temperature of the solid by one degree as it takes to raise the temperature of the gas: $3R$ instead of $3R/2$ per gram-atomic weight.

Because the thermal behavior of a gas made of independent atoms not connected together in molecules is relatively simple, those gases (helium, for example) are used as standards of temperature. Other kinds of thermometers, made of liquids or solids, are calibrated by comparison with a gas thermometer. Indeed, temperature is sometimes defined as a quantity proportional to the average kinetic energy of an atomic gas.

You might think that you could equally well define the temperature as a quantity proportional to the average kinetic energy of an atomic solid. The foregoing description of solids suggests that any temperature scale based on the behavior of a gas could be based just as securely on the behavior of a solid, except for a constant factor of 2. But the thermal behavior of a solid is actually much more complicated, as the next chapter shows.

III. HEAT CAPACITY

Nature certainly seems to move in jerks,
indeed of a very definite kind.
MAX PLANCK, *A Survey of Physics*

TWO IMPORTANT facts about the heat capacities of solids still escape the clutches of the reasoning in the last chapter. In the first place, atomic solids such as boron and carbon have much lower heat capacities than the rule of Dulong and Petit would suggest. In the second place, the heat capacities of all solids turn out to depend on the temperature at which they are measured. If Dulong and Petit had lived in a very much colder world, they might never have discovered their law.

The result derived in the last chapter gives a heat capacity of 6 calories per gram-atomic weight for any atomic solid, whether the solid is heated from 10 to 11 degrees or from 1000 to 1001 degrees. In fact, however, the heat capacity of any solid is lower at low temperatures than at high temperatures and even approaches zero near absolute zero (−273 degrees centigrade).

The resolution of these difficulties by Albert Einstein in 1907 was one of the earliest triumphs of the quantum theory, developed at the turn of the century by Max Planck. Surprisingly, the variation of the heat capacity with the nature of the solid and also with the temperature can be explained at one stroke.

Quantum-Mechanical Oscillators

Planck was led to the quantum theory by his efforts to explain the properties of the light radiated by a black body when it is heated. A hole leading into an empty, opaque cavity furnishes a very black body. When the cavity is heated, light of many different wavelengths streams from the hole. The light originates in the vibrations of the atoms at the inner surface of the cavity. Moving rapidly back and forth and carrying an electrical charge with it, each atom acts like a little radio antenna, radiating electromagnetic waves.

Some waves have very short lengths, which fall in the visible region, and produce a reddish glow; others are longer, falling in the infrared region; others, still longer, have the lengths of radio waves. Many experimental studies have been made to determine what fractions of the radiant energy are carried by waves of various lengths. Planck found that he could explain these experimental results only by assuming that each of the radiating atoms must emit or absorb energy in definite little amounts and not continuously.

Planck was able to show that if each radiating atom is vibrating like a harmonic oscillator, it must gain or lose energy in discrete energy bundles. Each bundle has the magnitude $h\nu$, where ν is the frequency—the number of times per second—at which the oscillator oscillates and h is a constant that has the same value for any oscillator, whatever its frequency. Planck's analysis of the black-body radiation fixed the value of that constant ($h = 6.62 \times 10^{-27}$ erg-second); it is now known as *Planck's constant*.

The novel idea that an oscillator can only be in discrete energy states separated by $h\nu$ could not have arisen from examination of balls and springs of visible size. It implies that any vibrating oscillator (a pendulum, for example) can choose its amplitude—its distance of travel—only out of a definite set of amplitudes. Figure 1 contrasts what the quantum theory would say about an ordinary pendulum with what it would say about a vibrating atom.

A pendulum 3 feet long, with a bob weighing $\frac{1}{2}$ pound, swinging 4 inches each side of the perpendicular, is oscillating with about $1/100$ foot-pound (about 10^5 ergs) of energy. The spacing of its permitted energy levels is an immeasurably small fraction of that energy: 10^{-32}. An atom, vibrating in a solid at ordinary temperature, may have its vibrational energy levels spaced by as much as 10^{-1} of its average energy of vibration.

Thus, the successive energy states of the pendulum would lie so close together that no experiment could show you the discontinuities between them. But the discontinuities in the permitted energy of an oscillating atom may be appreciable fractions of its total energy. In determining the properties of a huge assembly of atomic oscillators, the cumulative effect of those discontinuities may be very great.

The harmonic oscillator is agreeably simple to use in models because its frequency is the same whatever its amplitude may be. Hence, the difference in energy between each energy state and the next permitted by the quantum theory is a constant that depends only on the frequency of the oscillator. The energies that the oscillator can have are, in order of increasing energy, E_0 (some lowest energy), $E_1 = E_0 + h\nu$, $E_2 = E_0 + 2h\nu$, and so on (Fig. 1).

Imagine now what might happen if the solid model of harmonic oscillators were slowly heated, starting at absolute zero, where all its oscillators are in their lowest energy state. As the temperature of its surroundings is raised, no oscillator can accept less heat energy than the finite quantum $h\nu$.

The whole solid, made of a large number N of oscillators, can gain the energy $h\nu$ in just N different ways: any one of the N oscillators can pick it up. Hence, there are N ways by which the solid can hold the tiny added energy $h\nu$. When a second quantum of energy $h\nu$ is added to the solid, so that it has $2h\nu$ all together, there

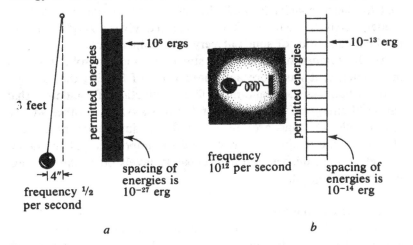

FIG. 1—Spacing of energies permitted (a) to an ordinary pendulum and (b) to an atom in a solid.

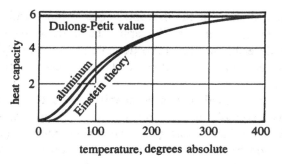

FIG. 2 – Measured and calculated heat capacities of aluminum.

are many more ways by which it can hold the energy. Any one oscillator might have $2h\nu$, or any two oscillators might have $h\nu$ apiece. Clearly, the number of ways by which the N oscillators can hold energy increases very rapidly as the energy increases.

The system of oscillators can pick up energy more easily as the number of ways in which it can hold the energy increases. Thus, at low temperature it takes a larger rise in the temperature to supply heat to the model than it would take if the oscillators would accept any amount of energy, however small. But that statement is exactly the same as the statement that a small amount of heat absorption corresponds with a large rise in temperature, or, in other words, that the heat capacity is small. In this way the harmonic-oscillator model, when the quantum theory is applied, gives a good picture of the heat capacities of solids at very low temperatures.

As the temperature increases still more and the solid acquires more energy, it has more and more ways of holding that energy. Thus, it behaves more and more like a collection of oscillators that can hold energy in an infinite number of ways, or, in other words, like oscillators that are unrestricted by any quantum effects. Hence, as the temperature rises, and the importance of the quantum effects diminishes, the heat capacity levels off at the Dulong-Petit value (Fig. 2).

Adjustable Parameters

The model even explains the small heat capacity at ordinary temperature in solids, such as boron and carbon, that have high melting points and are made of light atoms. The high melting point suggests that the atoms are exerting especially strong restraining forces on one another. Correspondingly in the model, the springs are unusually stiff. Moreover, the low atomic weight of the atoms makes the mass of each oscillator unusually small. Small mass and great stiffness are exactly the conditions for making the frequency ν of an oscillator unusually large. But an unusually large value of $h\nu$ makes an unusually large difference between the levels of energy permitted to the oscillator. That larger energy difference prevents the model from yielding the Dulong-Petit value of heat capacity until it reaches higher temperatures.

In short, Einstein's model of a solid accomplishes two things. First, it always gives the observed values of heat capacity at two extremes of temperature – a value approaching zero near the absolute zero of temperature and the Dulong-Petit value at a very high temperature. Second, by assigning a different frequency to the oscillators, the model is able to take account of differences in the behavior of different materials between those two extremes. Since Einstein's formula for the specific heat contains ν, a value for the frequency can be found that will make the formula fit best to the curve of heat capacities actually measured for a material. For example, the values of ν found in this way for aluminum and for lead are as follows:

Aluminum	6.6×10^{12} cycles per second
Lead	2.2×10^{12} cycles per second

A number that serves such a purpose is called an *undetermined parameter* by scientists, and they will look at it rather sharply. A formula can be cooked up to fit any experimental data by putting enough undetermined parameters in the formula. In a model designed to approximate the physical world, it is important that each parameter should have some sound physical basis. In the case of the frequency ν, the question arises whether, when the best value of ν has been chosen to fit a curve of measured heat capacities, that value makes any sense as a real frequency of vibration for a real atom in a real solid.

Fortunately, there are ways of showing that the very high frequencies in the table for ν make good physical sense. One way is to study the behavior of light passing through solids, as will be ex-

plained in a later chapter, after we look in more detail at how solids
are constructed of atoms.

First, however, it is worth while to examine a feature of Einstein's
model that still makes the calculated values depart somewhat from
the experimental values of heat capacity, no matter what value may
be assigned to the frequency ν. In accomplishing its major purpose — to explain the lower values of heat capacity at low temperatures — Einstein's model goes a little too far. As you saw in Fig. 2,
the calculated heat capacity approaches zero too rapidly at low
temperatures; the model overcorrects somewhat the error of Dulong and Petit.

Coupled Oscillators

This overcorrection sets a new problem: How can Einstein's
model be reasonably modified in a way that raises its heat capacity
at low temperatures? The success of the model in adapting itself to
boron and diamond immediately suggests one way. The model

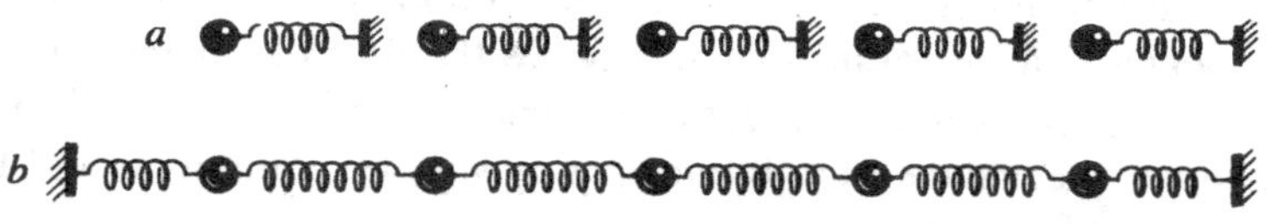

FIG. 3 — A one-dimensional solid represented (a) by a model in which the
atomic oscillators are independent and (b) by a model in which the atomic
oscillators are coupled to one another.

accommodates those two elements in their solid form by assigning
a high frequency to their oscillators and thus making it more difficult for the oscillators to accept heat. Hence, conversely, if a very
low frequency is assigned to some of the oscillators, those oscillators will be able to absorb heat more easily at low temperatures.
They will raise the heat capacity where it needs to be raised, while
the remaining high-frequency oscillators will continue to restrain
the rise of the heat capacity at higher temperatures.

At first sight this way of handling the difficulty looks unpleasant.
It is hard to see how to prevent at least two more undetermined
parameters from appearing: (1) a second frequency and (2) the
fraction of oscillators having that frequency. When the atoms are
all alike, it seems quite unreasonable to divide them into classes

represented by oscillators widely different in frequency. Nevertheless, in 1914 Peter Debye found a way to do this – a way so reasonable that it provides a more realistic model than Einstein's.
Debye's model still contains only one undetermined parameter and
can reproduce with amazing fidelity the heat capacities measured
over a wide range of temperatures.

Debye recognized that the oscillators are not oscillating independently. In fact, each oscillator is intimately coupled to its immediately neighboring oscillators all the time. If the solid were a
one-dimensional thing – a line of atoms – instead of a three-dimensional object, you could think of it as being more like the coupled

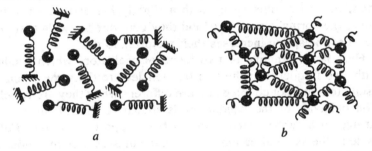

a   b

FIG. 4 – Coupling the oscillators to transform model a into model b, the
interconnected model visualized by Debye.

oscillators in Fig. 3b than like the independent oscillators in Fig.
3a. In two dimensions, the model shown in the last chapter (Fig.
4a) would be replaced by that in Fig. 4b. The mass-points would be
connected by springs to their neighboring mass-points, not to
supports outside the solid.

Now consider some of the interesting properties of coupled
oscillators. First imagine what will happen if we take two identical
pendulums and couple them by a very weak spring (Fig. 5a). If we
pull them both in the same direction – each the same distance from

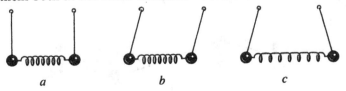

a   b   c

FIG. 5 – Two identical pendulums coupled by a spring.

its rest position—and then let them go (Fig. 5b), they will still swing with the same frequency, carrying the spring between them and feeling no force from it, because they neither expand it nor compress it.

But if instead we pull the pendulums in opposite directions and let go (Fig. 5c), the spring expands and adds its force to the gravitational force tending to pull the pendulums back toward their rest positions. They swing toward those positions a little more rapidly than before, and after they have passed through those positions, they begin to compress the spring so that it again adds its force to the gravitational force urging the pendulums back. In this case, then, the pendulums swing with a slightly higher frequency than when the spring is not there, and they continue to swing that way, because the spring increases their restoring force.

Here is an example of two identical oscillators which oscillate with two frequencies: the two frequencies are the same when the oscillators are not coupled and are different when they are coupled. The behavior of the pendulums is similar to that of the masses, strung on a rubber band between two supports (Fig. 6a). Pulled aside in the same direction (Fig. 6b), the masses oscillate about the center line at one frequency; pulled aside in opposite directions (Fig. 6c), they oscillate at a higher frequency.

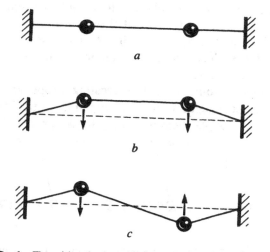

FIG. 6—Two identical masses strung on a rubber band.

These examples have shown only a higher, not a lower, frequency for the coupled oscillators. But lower frequencies appear as soon as more than two masses are coupled. It is easier to see how this will happen by examining masses on a rubber band than by examining pendulums. For example, the oscillations of the three masses shown in Fig. 7 might be started in three different ways.

Started in one way (1), the center ball will stand still and the other two balls will oscillate with the same frequency as the separate oscillators shown at the left side of the figure. Started in a second way (2), the oscillation will have a higher frequency, because the oscillations of the center ball help to pull the other two balls back, and vice versa. Started in the way shown in line (3), however, the oscillation will have a frequency lower than that of the separate oscillators. The angle between the two lengths of rubber band at any of the balls will be less than the corresponding angle for the independent oscillators, and therefore the restoring force and the frequency will be lower.

Stretched Strings and Elastic Vibrations

Notice that in the case of two coupled pendulums two frequencies appeared; in the case of three balls on a rubber band, three frequencies. Similarly, a line with a great many masses, like the model of a one-dimensional solid suggested in Fig. 3b, can oscillate

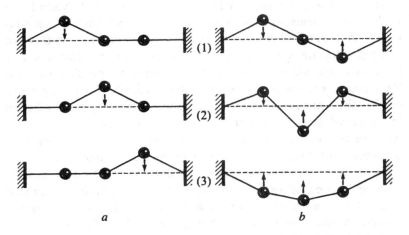

FIG. 7 — Three identical oscillators (a) uncoupled and (b) coupled

with a great many frequencies — the same number of frequencies as the number of masses. Some of the frequencies will be higher and some lower than the frequency of the uncoupled masses. In the eighteenth century the French mathematician Adrien Legendre worked out the different ways in which a stretched string, such as a violin string, could vibrate by imagining it to be made of an infinite number of masses, each separated from its neighbors by an infinitesimal distance and coupled to them.

As everybody who has learned a little about musical instruments will remember, a stretched string can vibrate in an enormous number of different ways, but there are some among that number that are especially simple and important. Those are the ways in which the string emits a pure tone. The tone of lowest pitch is often called the *fundamental* and the others are called its *harmonics*. The shapes the string takes when it is emitting the fundamental tone and its first five harmonics are shown in Fig. 8a.

Pitch and frequency are substantially the same thing: the lower the frequency of vibration of the string, the lower the pitch of the tone that you hear. Raising the pitch of a tone by an octave is equivalent to doubling its frequency. A fundamental tone and its first five harmonics are equivalent to a fundamental frequency and frequencies two, three, four, five, and six times that fundamental frequency.

Figure 8b suggests how Legendre's procedure would yield these well-known results for a stretched string. The six ways of oscillating six masses coupled in a line bear a conspicuous resemblance to a string vibrating in its fundamental and its first five harmonics. For the string, vibrations of still higher frequency are possible because the string is continuous, while no more than six elementary kinds of vibrations are possible for six masses. Of course, the number of independent sorts of vibrations of the string must be finite too, because the string is composed of a finite number of atoms — an enormous number, to be sure, but nevertheless a finite one.

Ideas like these were in Debye's mind as he set out to modify Einstein's model of a solid to improve the representation of specific heats. In imagination he replaced Einstein's identical uncoupled oscillators with the same number of coupled oscillators. He assigned to one of them the frequency of the fundamental vibration of

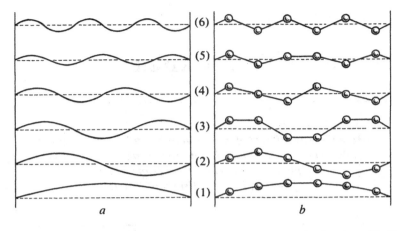

FIG. 8 — (a) The first six modes of motion of a stretched string. (b) Six coupled masses evenly spaced along a line.

the whole block of solid. He assigned to the host of remaining oscillators the frequencies of the harmonics of that fundamental, beginning with the lowest and proceeding upward until each oscillator was tagged with some frequency. The highest of all these frequencies then became the one undetermined parameter in the theory.

The harmonics of the fundamental frequency in a three-dimensional solid are not quite so simple as those in a one-dimensional string. Nevertheless, they can be determined approximately if you think of the solid as a structureless lump of matter — a three-dimensional analog of a continuous string — and use the results of the theory of elasticity, mentioned in Chapter I. Debye's theory is a good example of how the macroscopic and atomistic approaches to the explanation of solids can supplement each other.

A theory that explains so convincingly the heat capacities of most solids gives us the right to speak of *normal behavior*. If the heat capacity of a particular solid departs from that behavior, the departure can be examined and the reason for it can be searched out. In later chapters, you will come upon two such departures: one gives insight into the behavior of the electrons responsible for conducting electric current through metals and the other into the remarkable properties of ferromagnets.

IV. ORDER

Here the wave theory of X rays and the atomic theory of crystals come together, one of those surprising events to which physics owes its powers of conviction.
MAX VON LAUE, *History of Physics*

AT LEAST a century ago, careful observers had recognized that most solids are *crystalline*. The fact is conspicuous in rocks, and in metals it becomes immediately apparent under a microscope. The crystals — sometimes large, sometimes very small — are jumbled together in disorder and packed tightly.

The crystals do not usually show the flat natural faces that you may associate with the word "crystal." They have grown in the molten rock or metal when it cooled to its freezing point and have stopped growing whenever they came in contact with their neighbors, finally leaving no liquid. But the noncrystalline appearance of these grains of matter belies their inner structure. Examination with more penetrating tools will show unmistakably that they possess the same properties as obviously crystalline polyhedra that the same materials will form when they solidify slowly, free from confusing obstructions.

X-Ray Diffraction

It was suspected for a long time that the symmetry of the forms adopted by freely growing crystals is due to an underlying orderliness in the arrangement of their component atoms. In 1912 this suspicion was confirmed in Max von Laue's experiments with crystals and X rays. He reasoned that if a crystal does consist of an orderly array of atoms, and if X rays are rays of light whose wavelength is unusually short, then a crystal should diffract X rays much as a regularly ruled grating diffracts ordinary light. He found, in fact, that a single X-ray beam is diffracted by a crystal into many beams, each at a definite angle to the original beam.

If the terms *diffraction* and *ruled grating* are unfamiliar to you, you can get an idea of what led von Laue to his investigation by performing a simple experiment. Hold a phonograph record at the level of your eye with a lamp or a sunny window beyond it, and tilt the record so that the light glances off it at a very small angle (Fig. 1). Looking at the nearer side of the record, you will see bands of color, perhaps separated by bands of darkness.

In that experiment the regularly spaced grooves in the phonograph record furnish a crude ruled grating. Like X rays, the rays of light striking the record are trains of electromagnetic waves. If those waves all had the same wavelength, the light would have a strong color—the color characteristic of the wavelength. Then the light reflected by the record to your eye would be spread into several bands, separated by darkness and all of the same color.

FIG. 1—Light diffracted by the grooves of a phonograph record.

In order to understand why this separation would occur, look at Fig. 2. You can see that two waves of the same wavelength, proceeding at the same velocity in the same direction, will reinforce each other if the crests of one coincide with the crests of the other and will cancel each other out if the crests of one coincide with the troughs of the other. In Fig. 3 some consequences of such constructive and destructive interference appear for the beams scattered by the grooves of the phonograph record in Fig. 1. Reinforcement enhances the diffracted beams at the angles x and z, and cancellation darkens the beam at the angle y. When the beam contains light of many colors, as white light does, then (except for the simply reflected beam x) the reinforcements and cancellations occur at slightly different angles for the different wavelengths.

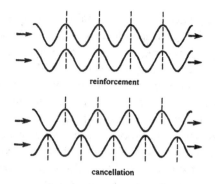

FIG. 2 – Reinforcement and cancellation.

The interference described in Figs. 2 and 3 can be expected only when the spacing of the successive grooves in the grating is not very different from the wavelength of the light. If the wavelength is hundreds of times shorter, the grooves are hundreds of wavelengths apart and become too crude an instrument to affect light in this way. Von Laue suspected that the waves of X rays were about the same length as the distances between atoms in a solid and that the orderly lines of atoms in a crystal might form a grating for the X rays.

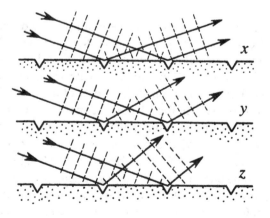

FIG. 3 – Diffraction of light by a grating.

To see more clearly how an orderly array of atoms scatters an incident beam of X rays into other beams, look first in Fig. 4 at how a single atom behaves in the presence of an advancing train of X-ray waves. Like light, X rays are electromagnetic waves, and as any electromagnetic wave passes the atom, it puts a rapidly oscillating electrical force on the atom. If the atom is electrically neutral, that force does not move the atom as a whole. But the atom is made of a positively charged nucleus with a cloud of negatively charged electrons around it. It is electrically neutral only because the positive charge on its nucleus is balanced by the negative charge on its electrons. At any instant, the electrical force of the X-ray wave pushes the nucleus one way and pushes the electrons the other. An instant later, the force reverses direction and pushes the nucleus and the electrons back again.

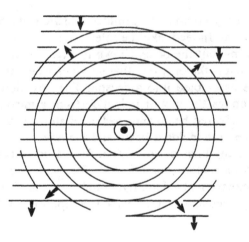

FIG. 4 — An atom driven by a train of X rays.

Thus, the X-ray wave produces an alternating current in the atom: the atom becomes an electrical oscillator. Like a tiny radio antenna, the atom radiates a wave synchronized with the X-ray wave that excites it. That reradiated wave proceeds outward in all directions from the atom as a train of spherical wave fronts.

In Fig. 5 you see how the spherical wave fronts from a large number of atoms add together when the atoms are evenly spaced

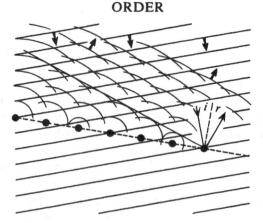

FIG. 5 – Reradiated X rays adding together to give a principal beam.

along a line. They combine to produce a plane wave front and thus a scattered beam. Indeed, like the ruled grating, they produce *many* scattered beams, but the beam shown in Fig. 5 is of special importance because it is the only one that survives when many such lines of atoms are assembled into a crystal. The direction of that scattered beam obeys the law of reflection from a mirror: the angle *r* between the reflected beam and the line of atoms is the same as the angle *i* between the incident beam and the line.

When the many lines of atoms that form a crystal are added, the analysis becomes more complicated. But von Laue was able to analyze his experiments similarly, confirming that X rays are electromagnetic waves and that a crystal is a repetitive orderly array of atoms.

Determining Atomic Arrangements

Following up von Laue's discovery, Sir Lawrence Bragg found a way of thinking about X-ray diffraction that greatly simplified the problem of interpreting the angles between the diffracted beams. He and his father, Sir William Bragg, applied *Bragg's law* to determine the atomic arrangements in many crystals and established X-ray diffraction as the central tool of modern crystallography. Others have progressively improved this tool and have worked out the atomic arrangements in a very large number of solid materials.

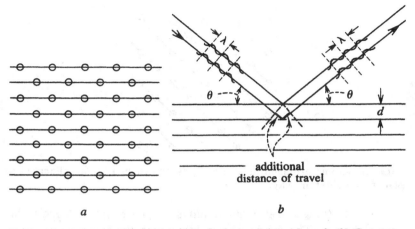

FIG. 6—(*a*) Regularly spaced planes of atoms in a crystal. (*b*) Constructive interference of X rays.

To pursue Bragg's suggestion for understanding X-ray diffraction in a crystal, think of successive similar planes of atoms, evenly spaced in a crystal (Fig. 6*a*) as if they were a succession of incompletely reflecting mirrors. Figure 6*b* shows an incident beam of X rays and one of the beams into which it is scattered.

When the X rays are all of the same wavelength, you can think of them in a way very similar to the one you used in Fig. 3. The fraction of the beam that is reflected by the top plane in Fig. 6*b* will be reinforced by the fractions reflected by the others only when the additional distance that those other fractions must travel is an exact multiple of the wavelength of the X rays. In other words, it is only then that a strong diffracted beam will be observed. This requirement can be put in the form of a mathematical relationship between the wavelength of the X rays, the spacing of the atomic planes, and the angle of the incident beam to those planes. The spacing d of the atomic planes, the wavelength λ of the X rays, and the glancing angle θ of the X-ray beam, are related by Bragg's law ($n\lambda = 2d \sin \theta$, where n is an integer) in any strongly diffracted beam.

The same crystal can be regarded as made of many differently chosen planes of atoms, at different angles, with different spacings (Fig. 7). The atomic arrangement is inferred by determining, with

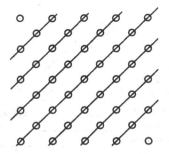

FIG. 7 — Atoms arranged in a crystal, as shown in Fig. 6*a*, allocated to planes in a different way.

the aid of Bragg's law, what families of planes give strongly diffracted beams. The inference employs much the same ingenuity as that required to solve a crossword puzzle.

This procedure has shown that almost all the materials that we call solid exhibit repetitive orderliness in their atomic arrangement. The form of this orderliness can always be described by repeating some pattern, containing relatively few atoms, again and again in space. Figure 8*a* shows an imaginary two-dimensional example of such a crystal structure, containing two species of two-dimensional atoms: white and black. In this example, the entire crystal can be built up by fitting together identical hexagonal units. Such a unit appropriate for building a particular crystal structure is called a *unit cell* for the structure.

Since a unit cell contains complete information regarding the arrangement of the atoms, the cell alone is often used to diagram a structure. You will hear the phrase, "the unit cell of" such-and-such a structure, and you may be tempted to infer that there is only one kind of unit cell that describes that structure. In fact, however, any structure can be described by an infinite number of different kinds of unit cells.

One way by which the choice of a unit cell can be varied is by moving it about (Fig. 8*b*). Another way is by changing its size (Fig. 8*c*). A third way is by changing its shape (Fig. 8*d*). The only requirement for a unit cell is that it build the structure correctly when replicas of it are set side by side so as to fill space. The unit cell is merely one of an infinite number of possibilities satisfying this requirement—a cell singled out because it provides the simplest description.

The Diversity of Order

The regular repetitive orderliness of a crystal is the distinguishing feature of a solid. In the vocabulary of physical science *solidity* and *crystallinity* are almost synonymous, and orderliness of solids accounts for many of their most familiar characteristics.

For example, the great diversity found among the orderly solids contrasts conspicuously with the great similarity found among the disorderly gases. Perfect disorder is a single thing, whereas perfect order is many things. Gases differ only in the chemical composition of the molecules that compose them. Solids can differ not only in their molecular species but also in the kinds of order that their molecules adopt.

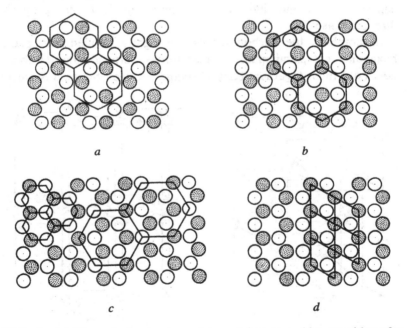

FIG. 8—(*a*) The orderly structure of a crystal portrayed by repetition of a single unit cell. (*b*) The unit cell after it has been moved to a new position. (*c*) Smaller and larger unit cells chosen for structure *a*. (*d*) A different shape of unit cell for structure *a*.

Indeed, the kind of order adopted by the molecules can be even more important than their chemical nature in determining the physical properties of the solid that they construct. You can see this clearly by examining instances in which a single substance can crystallize with two different atomic arrangements. A dramatic example is carbon, which occurs in two crystal structures, diamond and graphite (Fig. 9). Atoms of one and the same element, carbon, can form either a hard transparent solid or a soft black solid.

In the diamond structure each carbon atom is tightly connected with four others. The bonds stand rigidly toward the four corners of a regular tetrahedron. You can visualize those bonds as forming a three-dimensional network that assists in giving to diamonds their extraordinary hardness. But the planes (Fig. 10) perpendicular to those bonds are crossed by a smaller number of bonds than are other planes through the network. For that reason the gem-cutter can cleave diamonds into octahedra by sharp blows on a properly directed chisel.

In graphite, the carbon atoms are tightly bonded into plane hexagonal nets, and much weaker forces perpendicular to those nets hold them together. Easy cleavage between the nets is partly

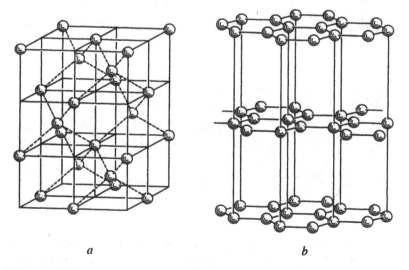

a   *b*

FIG. 9—Different arrangements of the atoms of carbon in (*a*) diamond and (*b*) graphite.

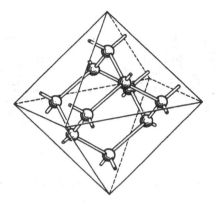

FIG. 10—Octahedral cleavage in diamond.

responsible for the utility of graphite as a lubricant.

There is an interesting relation between these crystal structures of carbon and two general classes of chemical compounds in which carbon is an important ingredient. The four tetrahedrally directed bonds formed by each carbon atom in a diamond are similar to the bonds formed by carbon atoms in molecules of the organic compounds classed as *aliphatic*. In the hydrocarbons whose molecules are shown in Fig. 11, each molecule is like a tiny segment of the diamond structure, terminating in hydrogen atoms.

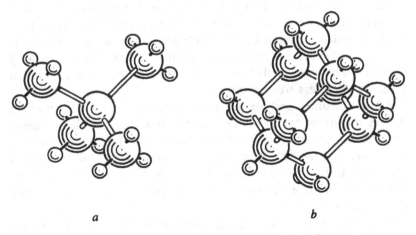

a   b

FIG. 11—Molecules of (a) neopentane and (b) adamantane.

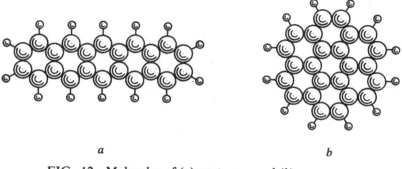

FIG. 12 – Molecules of (*a*) pentacene and (*b*) coronene.

On the other hand, carbon atoms are bonded to one another in plane hexagonal rings in molecules of the hydrocarbons classed as *aromatic* (Fig. 12). Here each molecule is like a tiny segment of one of the plane nets in the graphite structure, terminating again in hydrogen atoms. You can think of a diamond as a single gigantic aliphatic molecule and of a graphite crystal as a stack of gigantic aromatic molecules.

Partial Order

The comparison of solid carbon with organic molecules points to another way of looking at the diversity of crystalline orderliness. It is like the diversity of molecular structure that makes one species of molecules different from another. But in solids the diversity is made especially evident by the repetitive nature of the orderliness. A gas of disordered methane molecules differs from a gas of disordered benzene molecules, to be sure, but not nearly so much as diamond differs from graphite.

Perfect disorder and perfect order have one property in common that is important to the success of the physical theories of gases and solids. They can both be specified with precision. Liquids, which are intermediate between gases and solids, do not enjoy this property. It has proved much more difficult to construct a satisfying physical theory of liquids than of gases, on the one hand, and of solids, on the other.

Liquids resemble solids in that their molecules are packed closely together, and they resemble gases in that the molecules are arranged in a disorderly way and can move past one another as they jostle in their perpetual thermal unrest. A study of X-ray diffraction in a liquid will show often that some of its molecules fall into orderliness in little patches and then into disorder again.

In glass these little patches become frozen. A glass consists of a mixture of disordered and ordered molecules, all tied in positions where they can vibrate but cannot move past one another. Thus, a glass is like a solid in its rigidity and like a liquid in its relative disorderliness.

This patchy orderliness, evanescent in liquids and frozen in glasses, is only one example of how widely crystallinity appears in nature. Natural and artificial fibers and the protein structures of animals exhibit repetitive atomic orderliness, and some viruses have been crystallized. The methods of crystallography have been found useful even in the biophysical inquiry into genetics and the nature of life.

Defects

If glass is regarded as a special kind of liquid—a "supercooled" liquid, as it used to be called—patches of permanent order are its distinguishing feature. On the other hand, an approach to glass from the side of the solids emphasizes its relatively greater disorderliness and suggests a search for other kinds of defects that afflict the orderliness of solids.

It turns out that solids are subject to many kinds of defects, of which glassy disorder is only one. The most conspicuous defects in any crystalline solid that does not consist of a single crystal are the *grain boundaries* between the crystals that are jumbled together in it. But even the constituent crystals usually contain more subtle defects within them.

Here and there in the orderly arrangement of atomic sites, a site may be left unoccupied by its appropriate atom, becoming the site of a vacancy (Fig. 13*a*). Here and there a site may be occupied by an inappropriate atom, thus accommodating a *substitutional impurity* (Fig. 13*b*). Sometimes enough space is left in an orderly structure to invite the entrance of smaller atoms as an *interstitial impurity* (Fig. 13*c*).

Substitutional impurities are especially important. Crystallization

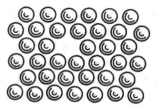

a

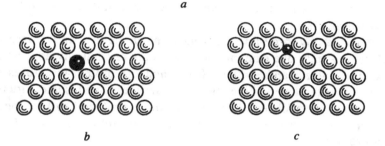

b   *c*

FIG. 13 –(*a*) Vacancies in the orderly arrangement of atoms in a crystal. (*b*) A substitutional impurity in a crystal. (*c*) An interstitial impurity in a crystal.

is a process by which almost any material can be purified. The orderliness adopted by the molecules of a substance when they collect in a crystal usually takes a form peculiar to that substance. Since the molecules of different substances ordinarily adopt different forms of crystalline orderliness, they cannot join one another indiscriminately. When a substance crystallizes out of a solution, for example, the impurities will remain in the solution and can be discarded.

But when two substances are made of molecules with closely similar chemical behaviors and geometrical shapes, their crystals may exhibit the same form of orderliness. Then they will often join with each other to build crystals having their common form of orderliness, in which molecules of one and the other sort occupy the orderly sites at random. Clearly two such substances cannot be purified satisfactorily by crystallization.

Examples of such *isomorphism* are most frequently found among substances whose molecules are identical except for one atomic constituent. When that constituent is chosen from among the species in the same group in the Periodic Table of elements, described in a later chapter, the molecules often form isomorphous crystals, in which either substance will play host to the other. Crystals of the mineral corundum, for example, are made of aluminum oxide. When a little chromium oxide is present in solid solution, the crystals acquire the red color that gives them commercial value as rubies.

Dislocations

In recent years a curious type of defect called a *dislocation* has captured much attention. Dislocations disturb the orderliness of a crystal in the neighborhood of lines through it. The easiest way to see what is meant by a dislocation is to look at a cross section of an imaginary crystal, taken in a plane perpendicular to the line of such a defect (Fig. 14).

You can imagine generating the defect by cutting the crystal part of the way through, inserting an extra plane of atoms, and closing the crystal again (Fig. 15). After the crystal has been closed, there is nothing unusual about the arrangement of the atoms near most of that plane. The structure is unusual only near the line along which the extra plane comes to its end.

Another way to imagine generating such a dislocation is closely related to the physical phenomena in which dislocations reveal their presence. Suppose that the top half of a crystal is pushed horizontally. Pushing hard enough might make the top half slip across the bottom half by one atomic distance (Fig. 16a). But it might not be possible to slip the entire top half. Then the line

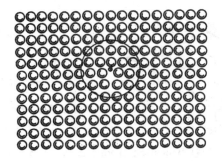

FIG. 14—An edge dislocation in a crystal.

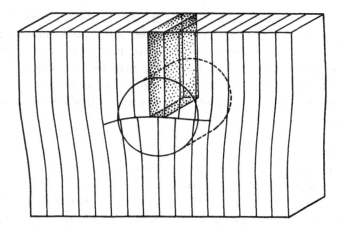

FIG. 15—Edge dislocation shown as the edge of an extra plane of atoms.

$D-D$, terminating the plane along which slip did occur, would be a dislocation (Fig. 16b).

Thinking a little further about this imagined way of generating a dislocation, you will see why dislocations are especially influential in the plastic flow of metals. Fortunately for technology, pieces of metal can be rolled and bent into useful shapes, which they retain after the fabricating forces have disappeared. Those forces achieve their purpose by making the constituent crystals slip along the planes of atoms that will move over one another most easily.

But most easily is not very easily. As in the cleavage of a diamond, the forces opposing the deformation, even in its easiest directions, are still large. In fact, if an entire plane of atoms tried to

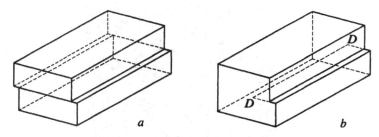

FIG. 16—Dislocation produced by a slip in a crystal.

slip at one time, the opposing forces would be so large that in order to be deformed, a metal would have to be pushed a thousand times harder than is in fact necessary.

The dislocations almost always present in large numbers in a metal crystal give it plasticity by enabling a plane of atoms to slip over a neighboring plane piecemeal. The deforming forces move the line of the dislocation through the crystal by moving first one row of atoms and then another.

Figure 17 shows four successive steps in this process. Driven by the shearing forces shown by the arrows, the region of misfit sweeps to the left. The two black atoms, on opposite sides of the plane of slip, shift their relative positions by one interatomic spacing as the dislocation moves past them. Thus, only a row, not a whole plane, of atoms slips at one time; but in the end the crystal has slipped by one atomic distance along the whole plane.

When a crystal has a "family" of planes of easy slip, along which dislocations can move readily, it may have several such families. For example, the structure used in this discussion of dislocations

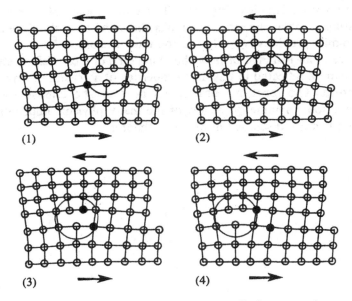

(1)   (2)

(3)   (4)

FIG. 17—How a dislocation assists slip in a crystal.

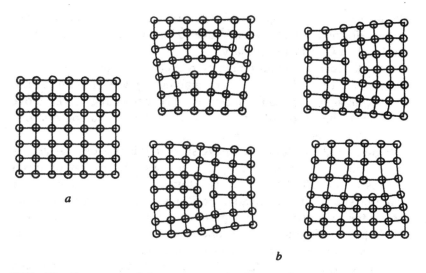

a

b

FIG. 18—Symmetry of the arrangement of atoms in (*a*) the ideal crystal in (*b*) four orientations.

has two such families (Fig. 18). The two families of planes are indistinguishable; the structure looks the same to both of them. If one family provides easy slip, so too must the other.

Here is an instance of how the symmetry of a crystalline structure can be reflected in physical properties of a crystal having that structure. The repetitive orderliness of a crystal almost always gives it some symmetry. A few of the many physical implications of that fact are discussed in the next chapter.

V. SYMMETRY

THE IDEA of symmetry is familiar but vague to nearly everyone. It provides a way of looking at many aspects of the world and becomes more and more helpful as it is made more precise. Crystallographers have achieved especially high precision in specifying the symmetries of crystals. Their methods not only illuminate solids but also suggest correspondingly precise means for using the idea of symmetry elsewhere. And as this chapter will show, the idea can even give crucial help in designing solids for practical uses.

You can distinguish macroscopic and atomistic approaches toward crystalline symmetry, as toward the rest of the physics of solids, and the two make an interesting contrast. From the macroscopic point of view, the symmetry of a crystal is the symmetry of its observable properties. From the atomistic point of view, the symmetry of a crystal is that of the arrangement of its atoms. Since all the observable properties of a crystal are contributed by its atoms, the two points of view must yield results that are consistent with each other.

The symmetry of a property is not so mysterious as it may sound. A crude example of the idea appears in the contrast between the strengths of wood and of steel. A piece of wood is much stronger in tension if the tension is applied along the grain than across the grain. The strength of a piece of steel, on the other hand, is more highly symmetrical; usually it has almost the same strength under tension in any direction.

At the end of the last chapter Fig. 18 portrayed a simple instance of how the symmetry of a property becomes consistent with the symmetry of an atomic arrangement. The symmetry duplicated at 90 degrees a family of slip planes.

The most conspicuous properties of some crystals are their shapes. When a crystal grows without running into obstacles, its shape is characteristic of its substance and of the form of orderliness that the molecules of its substance adopt. All crystals of the same substance grown without obstruction under the same conditions will have the same shape. That self-imposed shape, therefore, tells something about the solid substance. Long before X-ray diffraction revealed the arrangements of atoms in solids, crystallographers studied those shapes — in particular their symmetries — for whatever information they might yield.

Polyhedra

The shape of a freely grown crystal is almost always a polyhedron: it is bounded by a finite number of flat faces, meeting in edges that in turn meet at corners. Crystals of the mineral fluorite, for example, often have the shape of a regular octahedron (Fig. 1a). Crystals of alum grown from solution in water have a shape in which the corners and edges of a regular octahedron are cut off by other faces (Fig. 1b). Surely you would describe these polyhedra as symmetrical.

Thinking of what you mean when you say that, you will conclude that you call them symmetrical because there are many directions from which they look the same. There are two procedures that you might use to find all those directions. You could set the polyhedron on a pedestal and walk around it, or you could keep your head still and turn the polyhedron. Both procedures would yield the same answer, and when you start to specify the symmetry in detail, it is more convenient to think of turning the polyhedron than walking around it.

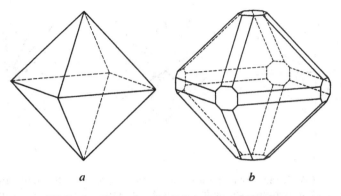

FIG. 1 — Crystals of (*a*) fluorite and (*b*) alum.

After you have turned the regular octahedron (Fig. 1*a*) 90 degrees around a vertical line through its opposite corners, it looks
the same; you cannot distinguish that you have done anything at
all. In fact, you can continue to turn the octahedron into indistinguishable positions in successive steps of 90 degrees. Including the
original position, four such positions appear in the course of a full
revolution, and the line about which the polyhedron turns is therefore called an *axis of fourfold symmetry*. Since the regular octahedron has six indistinguishable corners, occurring in three pairs, it
has three axes of fourfold symmetry (Fig. 2*a*).

In a similar way you can find four axes of threefold symmetry,
each of which passes through the centers of a pair of opposite faces
(Fig. 2*b*). Furthermore, through the center of each of the six pairs
of opposite edges there is an axis of twofold symmetry. In other
words, the octahedron is indistinguishable after it has made a
half-turn about any of the six lines shown in Fig. 2*c*, or one-third of
a turn about any of the four lines shown in *b*.

Here is the beginning of a precise way of specifying the symmetry of an object. Find all the different actions that can possibly be
performed on an object that would carry it into indistinguishable
positions. Each of those actions — for example, a rotation by
90 degrees about a line through opposite corners of an octahedron
— is called a *symmetry operation* for the object. Such things as the
line about which the rotation is performed are called *symmetry ele-*

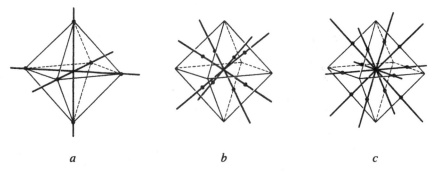

FIG. 2 — (*a*) Three axes of fourfold symmetry passing through opposite corners of a regular octahedron. (*b*) Four axes of threefold symmetry passing through opposite faces of a regular octahedron. (*c*) Six axes of twofold symmetry passing through opposite edges of a regular octahedron.

ments of the object. The entire collection of symmetry operations or of symmetry elements completely specifies the symmetry of the object.

Nonperformable Operations

You are halfway to that goal when you have found all the axes of symmetry. To complete the specification, some operations of another kind must be examined. This kind of operation differs from rotation in a fundamental way. You cannot physically perform operations of the second kind, as you can perform a rotation; you only can imagine performing them.

The most important of these operations is reflection. Imagine passing a plane through the object to divide it into two parts, and imagine that each side of the plane forms a mirror. If each part of the object looks like the image of the other part in the mirror, you can say that the object is unchanged by a reflection in the plane that you have chosen. Such a *plane of reflection symmetry* is the only symmetry element possessed by many living organisms (Fig. 3).

The octahedron has nine planes of reflection symmetry as well as its axes of rotation symmetry. Three of those planes are of the type shown in Fig. 4a, each passing through four of the twelve edges. The other six planes are of the type shown in Fig. 4b, each passing through two opposite corners and the centers of two opposite edges.

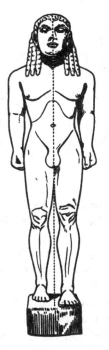

FIG. 3 – A plane of reflection symmetry as shown in the Argive Statue from Delphi. (From Herbert Green Spearing. *The Childhood of Art or the Ascent of Man*, London, Ernest Benn, Ltd., 1930. Used by permission.)

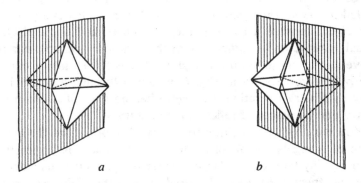

a   *b*

FIG. 4 – (*a*) Three planes of reflection symmetry, each containing four corners, passing through a regular octahedron. (*b*) Six planes of reflection symmetry, each containing two corners and the centers of two edges, passing through the regular octahedron.

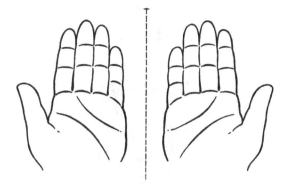

FIG. 5—How the nonperformable operation described by a plane of reflection would turn a right hand into a left hand.

Having visualized the operation of reflection in a plane by imagining that the plane is a two-sided mirror, you will understand a more precise description of that operation. Reflection moves every point in the object perpendicularly to the plane of reflection, through it, and out again to an equal distance on the other side of it. Figure 5 shows that reflection turns a right hand into a left hand and vice versa, and that fact makes especially clear the sense in which the operation is *nonperformable*.

There is another nonperformable operation, closely allied to reflection, that is a symmetry operation for many objects. It is called *inversion through a center*. Inversion moves every point in the object through a single point, the center, and out again to an equal distance on the other side of the center. For example, a right glove can be turned into a left glove if it is turned inside out. For the polyhedron shown in Fig. 6 inversion through a center is the only symmetry operation; the figure has no axes of rotation symmetry and no planes of reflection symmetry.

You are now in a position to say much more about the appearance of a regular octahedron than simply that it is symmetrical. You can say that it has a large number of symmetry elements, and you can specify them as 3 fourfold axes, 4 threefold axes, 6 twofold axes of rotation, 9 planes of reflection, and a center of inversion. You may be interested to verify that the alum crystal shown in Fig. 1*b* has the same symmetry.

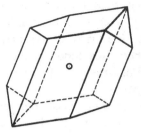

FIG. 6 — A center of inversion as the only symmetry element.

Atomic Arrangements

When a crystal grows naturally into an octahedron, as fluorite does, or into the shape shown in Fig. 1*b*, that symmetry must embody some symmetrical features of the crystal's inclinations. Since those inclinations are determined by the crystal's constitution, their source must be sought in the atomic arrangement.

The first step in examining the symmetry of a crystalline atomic arrangement is to recall that it has the property of repetitive orderliness. You can then imagine that the arrangement is repetitive in the fullest sense of that word: it is repeated in all directions in space to an indefinite distance.

Any particular crystal comes to an end, of course. Indeed, the way in which it comes to an end to form a polyhedron has just now been the subject of your examination. But each of the repeated unit cells in the crystal is so small that the crystal contains many millions, and the end of the crystal is many thousand unit cells away from almost any one of them.

The repetitive character of the orderliness in a crystal, extended indefinitely, introduces a new sort of symmetry operation for atomic arrangements, *translation*. A crystalline atomic arrangement can be moved bodily, without being turned or reflected, into new positions that give it an appearance indistinguishable from its old appearance. After such a translation all atoms are set down in positions that were previously occupied by exactly similar atoms.

Like polyhedra, atomic arrangements also can have axes of rotation symmetry and planes of reflection symmetry. Since the atomic arrangement is repetitive in a crystal, those axes and planes are repeated also; they pass through the same parts of each unit cell. You can think of a crystalline atomic arrangement (Fig. 7a) as possessing a repetitive array of symmetry elements (Fig. 7b). In that figure the dashed lines denote planes of reflection symmetry and the boat-shaped figures denote axes of twofold symmetry perpendicular to the paper.

Now you can see the connection between the symmetry of a polyhedron and the symmetry of the atomic arrangement in a crystal that grows into that form. The flat crystal used as an example in Fig. 7 might grow into the polyhedron shown in Fig. 8. The symmetry elements of the polyhedron are the same as those of the atomic arrangement but all passing through one point, the center of the polyhedron.

Properties

You can think of that center as the point at which the crystal started to grow. Some small group of atoms coming together in the right sort of orderliness formed a seed on which successive layers of atoms laid themselves in the same sort of orderliness. As the crystal grew, it retained a characteristic polyhedral shape that became progressively larger. A small regular octahedron, for example, would grow into a large regular octahedron.

Thinking now about that process of growth, you can make a connection between the shape of a crystal — a geometric property — and the growth rate of that crystal — a dynamic property. If a crystal grew from its seed at an equal rate in all directions, it would take the form of a sphere (Fig. 9a). Since no crystal naturally takes that form, you can infer that no crystal exhibits the same growth rate in all directions. By drawing arrows from the center of a crystal perpendicular to its faces (Fig. 9b), you can find the relative growth rates in those directions.

The diagram of arrows has the same symmetry as the polyhedron from which it was derived. You can speak of the symmetry of the growth rate of the crystal in the same way that you can speak of the symmetry of its shape. This suggests that any other dynamic property of the crystal that is capable of varying with direction will also have symmetry. If some means could be found for diagramming how the property varies with direction, the dy-

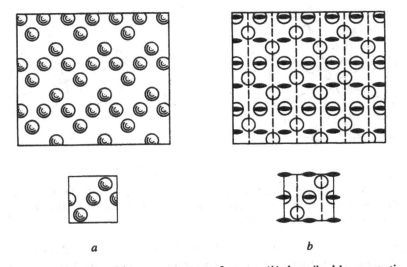

a   b

FIG. 7—(a) A repetitive arrangement of atoms, (b) described by a repetitive array of symmetry elements.

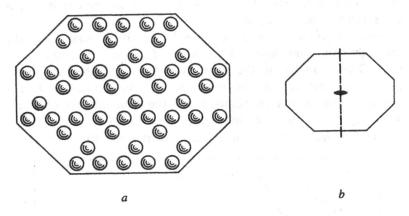

a   b

FIG. 8—(a) A crystal and (b) its symmetry elements.

namic symmetry of the property would be the same as the geometric symmetry of the diagram.

The conduction of heat in a crystal is a good example. If a slab is cut from a single crystal of quartz, coated with wax on one side, and heated at a point, the wax will melt away from that point into

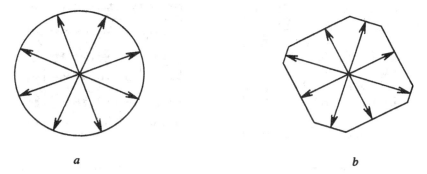

a

b

FIG. 9—(a) A crystal growing at the same speed in all directions. (b) Perpendiculars from the center to the faces proportional to the rates of growth in different directions.

an elliptical pool (Fig. 10). If the heat conduction were the same in all directions, the pool would be circular. But the heat conduction in a quartz crystal varies with direction; it is greatest along the direction of the long axis of the ellipse. In the directions contained in the plane of the slab, the symmetry of that property of quartz is the symmetry of an ellipse.

You will get an ellipse of the same shape, with its long axis in the same direction, no matter what point of the slab you choose to heat. The property is the same and has the same symmetry throughout the crystal. In other words, the crystal is *homogeneous*—the same at all points. But at the same time the crystal is *anisotropic*—not the same in all directions.

The homogeneity is a consequence of the fact that the atomic

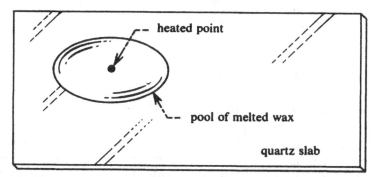

FIG. 10—Anisotropic conduction of heat in a quartz crystalline plate.

arrangement is repeated within very short distances, comparable to the size of a molecule. The crystal is not the same at all points if you look at it in the atomistic way, because there may be atoms of one kind at some points and of another kind at others. But you cannot see that variation directly in any macroscopic experiment; homogeneity is a macroscopic idea.

The anisotropy is a consequence of the fact that the atomic arrangement does not look the same from every direction, and the regular repetition of that arrangement has a cumulative effect on the bulk properties of the crystal. Since the crystal is the same at all points, when you look at it macroscopically, its macroscopic symmetry is also the same at all points. The symmetry elements appropriate to any macroscopic point in it are the same everywhere. They are the symmetry elements of the atomic arrangement but all passing through one point.

In other words, the symmetry of a crystal's macroscopic properties is related to the symmetry of its atomic arrangement in the same way that the symmetries of the polyhedron and the atomic arrangement in Fig. 8 are related. But the symmetry of its properties applies to all macroscopic points in the crystal, not just to the one point at the center of the polyhedron.

Of course, in a solid made of many little crystals jumbled together at random—in most pieces of metal, for instance—the anisotropy of each component crystal is hard to uncover. The directional differences within each crystal tend to average out in the solid as a whole, and in most macroscopic experiments the solid appears to be *isotropic*; that is, its characteristics are the same in all directions.

Causes and Effects

When you extend the idea of symmetry still further, some interesting kinds of physical reasoning about solids come to light. It is often convenient to think of physical phenomena in terms of influence and result—of cause and effect. Almost anyone applying a force to an object and observing that the object picks up speed will think of the force as a cause and the acceleration as an effect. And often you can usefully go on to think of both a cause and its effect as having characteristic symmetries.

For example, a force is completely described by its size and its direction. You can diagram that specification by an arrow: the length of the arrow represents the magnitude of the force, and the direction of the arrow represents its direction. The symmetry of a force is therefore the same as the symmetry of an arrow. The line of the arrow is an axis of symmetry for all angles of rotation, and all planes containing that line are planes of symmetry. No axes or planes of symmetry are perpendicular to the arrow, because its head is distinguishable from its tail.

For a simple example of how to use such ideas about a cause, examine the *dielectric polarization* of a crystal, mentioned in Chapter I. It will help you to paint a clear picture in your mind if you first look at an instance of dielectric polarization in an atomistic way and then switch to the macroscopic way of looking at the same problem.

Imagine that there are electrodes on the surfaces of a slab of the crystal whose arrangement of atoms is shown in Fig. 11. Imagine that each white atom carries a positive electric charge and that each black atom carries negative charge. An electric battery connected to the electrodes puts an electromotive force on the charged atoms, which pushes the positively charged atoms in one direction and the negatively charged atoms in the opposite direction.

If the atoms were completely unimpeded, they would rush to the electrodes. But they are packed together and feel forces from one another when they try to move. The electromotive force can push

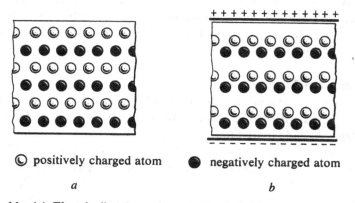

© positively charged atom   ● negatively charged atom

a   b

FIG. 11—(a) Electrically charged atoms in a crystal moved slightly by (b) an electrical force.

the atoms only a short distance before they stop. The motion of the electrically charged atoms through that short distance appears as a brief pulse of current called the *displacement current.*

In this example, you can think of the electromotive force as a cause and the displacement current as its effect. A current, like a force, can be completely specified by its magnitude and its direction; again, therefore, an arrow can represent it. In other words, here is an example in which an arrow (the cause) gives rise to another arrow (the effect) in a medium (the crystal). The reasoning that you are about to examine uses the symmetry of an arrow and the symmetry of the medium to find some necessary relationships between the directions of the two arrows. Then the same relations must apply to the directions of the electromotive force and the displacement current.

Cause, Effect, and Medium

Continuing for a moment the atomistic examination of the crystal in Fig. 11, you will notice that an electromotive force applied in the direction shown in Fig. 12a should move the atoms in the same direction. Because the arrangement of atoms is symmetrical around that direction, they will not put forces on one another that would move them in some other direction.

When an electromotive force is applied in the direction shown in Fig. 12b, however, the atoms need no longer move in the direction of that force. Their motion will take some direction determined in a complicated way by a combination of the applied electromotive

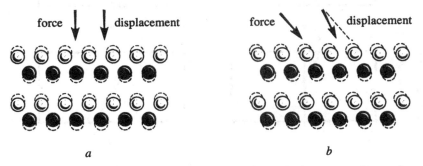

FIG. 12—Directions in which the charged atoms in a crystal may be displaced by an electrical force.

force, of known direction, and the many forces between the atoms, whose directions are unknown.

Now turn to look at how the same directional reasoning can be carried out in the macroscopic way. The symmetry of the crystal before application of the electromotive force is shown in Fig. 13a and b. The effect of applying the force can be represented by imagining that an arrow has entered the crystal (Fig. 13c and d). Thereupon the crystal loses much of its symmetry because its atoms move. In fact, the only symmetry remaining to it is that which is common to the undisturbed crystal and the disturbing arrow.

If the arrow lies in one of the planes of reflection symmetry of the crystal (Fig. 13c), that symmetry element is undisturbed, because the arrow has a plane of symmetry coinciding with it, but the other two planes of symmetry disappear. If the arrow lies in some other direction (Fig. 13d), all the planes of reflection symmetry are destroyed; no symmetry remains so long as the arrow is there.

Now examine the displacement current that results in these two cases by thinking of it as a property of the new medium that contains the arrow. Another arrow will represent that property. But if the new medium still has a plane of reflection symmetry (Fig. 13c),

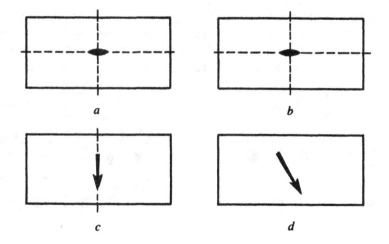

FIG. 13—Macroscopic symmetries of the crystalline material of Figs. 11 and 12, (a and b) without and (c and d) with applied electrical forces.

any arrow representing a proposed displacement must be duplicated by another, its reflection in the symmetry plane (Fig. 14a). The
resultant of the two displacements lies along the direction of the
arrow representing the force (Fig. 14b).

In this way you reach the same conclusions that you reached by
atomistic reasoning in Fig. 12a. When no symmetry remains to the
new medium (Fig. 14c), this reasoning yields no information, but
neither did the atomistic reasoning, so far as it was carried in Fig.
12b. In other words, when you do not know the atomic arrangement in a crystal but only its macroscopic symmetry, a macroscopic argument will often serve just as well as an atomistic argument, if both of these arguments depend only on symmetries.

The principle used in such arguments was first put in general
form by Franz Neumann in the last century. Neumann's principle
says that the symmetry elements of a medium in the presence of a

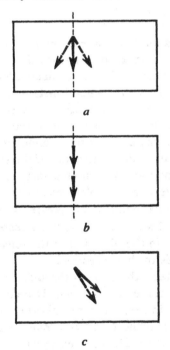

FIG. 14—Requirements of symmetry applied to proposed electrical
displacements.

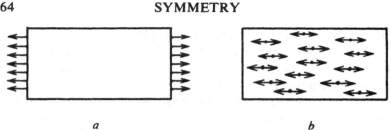

FIG. 15 — A tensile stress applied to a solid body.

cause can be no less than those that are held in common by the medium in the absence of the cause and by the cause in the absence of the medium. Since the effects of a cause are properties of the new medium, those effects must obey the symmetry of the new medium.

Stresses

Arguments of this kind are called *symmetry arguments;* they can often yield answers to physical questions that are much less trivial than this simple example. The piezoelectric effect, already mentioned in Chapter I, furnishes a good instance.

In piezoelectricity the effect is again an electrical displacement, but the cause is not so simple as an electromotive force. The cause is a stress, not a force, and a stress cannot be represented by a single arrow. In order to discover some of the symmetry properties of any stress, look first at a few simple sorts of stresses.

One of the simplest imaginable stresses is tension. To exert a tensile stress on a body, equal and opposite forces are applied to its two ends, as in Fig. 15a. Usually the body responds to the tension by elongating a little in the direction of the tension and contracting a little at right angles to that direction.

Clearly the body feels the tension throughout its bulk, not just at the ends where the forces are applied. It behaves as if each little part of it were tugging on each neighboring little part. You can think of the tension in the body as a pair of equal and opposite forces, working against each other everywhere, and therefore you can diagram it at every point by a pair of equal and opposite arrows (Fig. 15b).

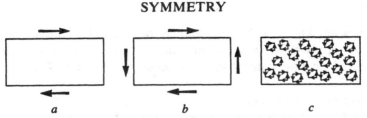

FIG. 16—A shearing stress applied to a solid body.

Another simple kind of stress is the shearing stress considered in the discussion of slip in the last chapter. Again you see a force couple—the pair of equal and opposite forces (Fig. 16a)—but the two forces are not in line with each other. In fact, they will make the body whirl around unless they are accompanied by another force couple (Fig. 16b). Again the stress is found throughout the body to which it is applied and could be represented by thinking of a little array of four arrows placed at every point in the body (Fig. 16c).

Examine in Fig. 17 the symmetries of these two kinds of stresses. You will notice that both have the same symmetry elements if you think only of symmetry operations that do not carry you off the plane of the paper. Those elements are two mutually perpendicular planes of reflection symmetry, an axis of twofold rotation symmetry, and a center of inversion. They are the same as the symmetry elements of an ellipse.

More complicated stresses can be imagined—stresses that cannot be represented by a collection of forces all lying in one plane. But it can be shown that even the most complicated stress can be

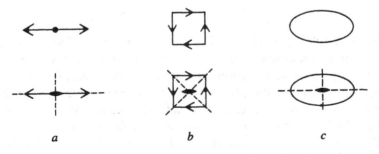

FIG. 17—Symmetry of (a) a tensile stress or (b) a shearing stress compared with that of (c) an ellipse.

represented by an ellipsoid and therefore has at least the symmetry of an ellipsoid: three planes of reflection symmetry, three axes of twofold rotation symmetry, and a center of inversion (Fig. 18).

A Symmetry Argument

Now ask yourself what materials can develop an electrical displacement when a stress is applied to them, and thus show a piezoelectric effect. Any isotropic material looks the same no matter how it is turned; all directions in it are axes of symmetry for rotation through any angle. Neumann's principle then says that when a stress is applied to such a material, it will retain as symmetry

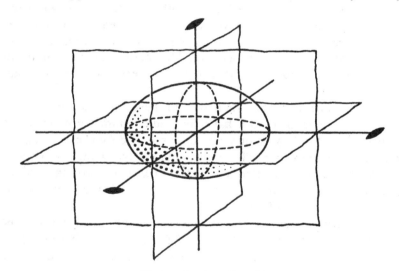

FIG. 18—An ellipsoid.

elements the 3 twofold rotation axes of the stress. The symmetry operations described by those axes will prevent the occurrence of any effect that can be represented by an arrow. One of the twofold axes requires that the solid arrow be accompanied by the dashed arrow (Fig. 19a). A second twofold axis requires that the pair in Fig. 19a be accompanied by a canceling pair (Fig. 19b). Hence, a stress cannot produce an electrical displacement in an isotropic material.

Turning from isotropic materials to crystals, we find many dif-

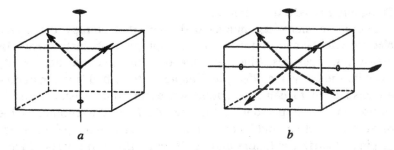

FIG. 19—A proposed displacement produced by a stress in an isotropic medium.

ferent types of symmetries. Each of those types differs from the symmetry of an isotropic material in the fact that the symmetry elements are finite in number. Hence, it is always possible to apply a stress in such a way that its three axes of rotation symmetry and its three planes of reflection symmetry do not have the same direction as the axes and planes in the unstressed crystal. Neumann's principle then says that the stressed crystal will retain none of its symmetry axes and planes.

But if the unstressed crystal has a center of inversion symmetry, it will retain that center even when it is stressed because the stress also has an inversion center. A medium having such a center cannot exhibit any effect representable by an arrow (Fig. 20). Hence, crystals with a center of inversion cannot exhibit the piezoelectric effect. On the other hand, crystals lacking an inversion center can always exhibit that effect for some form of stress.

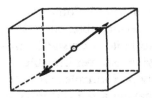

FIG. 20—A proposed displacement produced by a stress in a medium having a center of inversion symmetry.

Designing Piezoelectric Materials

There are many technological devices that employ the piezo-electric effect — microphones, phonograph pickups, instruments for measuring pressures. Since any piezoelectric material necessarily exhibits the converse effect, as Chapter I described, such materials are also used to generate sound waves in sonar systems and in equipment for hypersonic cleansing. And both phenomena — the piezoelectric effect and its converse — cooperate in the performance of piezoelectric oscillators and in elements that form parts of electric wave filters, vital to multiplex telephony. In one form or another, the symmetry arguments of this chapter furnish the best means for attacking the problem of finding or making the piezoelectric materials used in these devices.

For some time piezoelectric materials were chosen from among the crystals — quartz in particular — that had been shown by crystallographic studies to lack a center of inversion symmetry. More recently, materials suitable for many purposes have been made by electrically polarizing ceramic aggregates containing barium titanate. In these ceramics, which are isotropic before their polarizing treatment, the application of a high voltage produces a dielectric displacement that remains after the voltage is removed. Thus, these mediums lose their center of symmetry. In the terms of the preceding discussion, they can be pictured as mediums having the symmetry of the arrow that represents the permanent dielectric displacement within them.

It is possible also to use a symmetry argument to direct the design of molecules that must form crystals lacking a center of symmetry. The argument can be put in the following form: If an individual molecule lacks not only a center of inversion symmetry but also any plane of reflection symmetry, then it can have only axes of rotation symmetry. In other words, its only symmetry operations must be performable operations. Putting many such molecules together to build a crystal is manifestly a performable operation. But carrying out performable operations on identical molecules having only performable symmetry operations can never construct an object having a nonperformable symmetry operation. A box of right-hand screws, no matter how they are jumbled or arranged, will always have right-handedness in it, distinguishing it from a box of left-hand screws. Then, since inversion through a center is a nonperformable operation, a crystal of such molecules cannot have a center of symmetry. In one experimental check of this argument, all of the crystals so constructed exhibited the piezoelectric effect.

VI. ATOMS AND IONS

The periodic law has shown that our chemical individuals
display a harmonic periodicity of properties
dependent on their masses.
DMITRI MENDELEYEV, *Faraday Lecture of 1889*

"THE PARTICLES of Bodies stick together by very strong Attractions." We have put aside Sir Isaac's Business as long as possible; it is time to come to grips with the problem of why atoms cohere into solids. The theory of their heat capacity gave no clue to the answer; it took the cohesion of the atoms for granted. So too did the description of the orderliness and symmetry of the atomic arrangements in crystals; it did not ask why the atoms adopt those arrangements.

When Newton charged experimental philosophy with this task, he had already fathered the law of universal gravitation between masses. Who can guess whether he felt tempted to suppose that the gravitational force between the masses of the atoms supplied the "strong Attractions" that he sought? If he did, he soon realized that gravitational forces were entirely too weak to suffice.

In recent years it has become clear that the forces which bind atoms together into solids are in all cases electrostatic forces. Gravitational forces are present, of course, but are negligible in comparison with the force of attraction between positive and negative electrical charges. Indeed, the forces binding atoms together in molecules turn out to be electrostatic also. Newton was right in guessing that the forces that he sought were responsible for "chymical Operations" as well as for the cohesion of solids.

But to say that the forces of cohesion are electrostatic is one thing, and to explain how atoms that apparently are electrically neutral can exert electrostatic forces is quite another. Fortunately, one of the commonest solids, common salt, provides an instance easy to understand.

Ionic Bonds

Common salt is sodium chloride, made of sodium atoms and chlorine atoms in equal numbers. When salt is dissolved in water, the solution conducts electricity very well. As a direct current passes through the solution from one electrode to the other, sodium moves to the negatively charged electrode (the cathode) and chlorine to the positively charged electrode (the anode). This movement is evidence that, in solution at any rate, the sodium atoms are positively charged and the chlorine atoms are negatively charged. Such charged atoms are called *ions*.

All the evidence suggests further that the sodium and chlorine atoms bear charges not only in the solution of salt but also in solid salt. For example, the arrangement of the ions in a sodium chloride crystal makes this idea seem reasonable. Each sodium ion is immediately surrounded by six chloride ions, and each chloride ion by six sodium ions (Fig. 1). Each element has attracted ions of the other element and has pushed its fellows away, as the attraction between opposite charges and the repulsion between like charges would lead you to expect. You can go on to infer that the whole crystal of sodium chloride is held together by the electrostatic attractions between its positive and negative ions.

You might think that the repulsions between the like charges would cancel the attractions between the opposite charges, leaving no net force to hold the solid together. But the attractions outweigh the repulsions because the charge on the nearest neighbors of each ion is of opposite sign, and therefore the forces between all nearest neighbors are attractions. These forces are stronger than the repulsions from the charges on the more distant neighbors because the electrostatic force between any two charges, like the gravitational force between two masses, varies inversely with the square

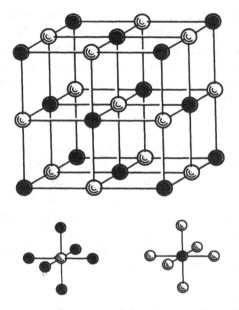

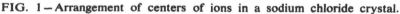

FIG. 1 — Arrangement of centers of ions in a sodium chloride crystal.

of the distance between the charges and consequently declines with increasing distance.

The inverse-square law of the force between two charges has been used to calculate how much work would be required to pull all the sodium and chloride ions apart from one another. When the calculated work is compared with measurements of the energy actually required to convert sodium chloride into a dilute vapor of its component ions, the agreement is very close. No question remains that solid sodium chloride is made of ions and that the electrostatic forces between those ions make them cohere.

But this explanation of cohesion raises questions that probe deeper: How and why do neutral atoms of sodium and chlorine acquire their electrical charges and become ions? The question "How?" can be answered quickly. Each chlorine atom takes an electron from a sodium atom. The extra negatively charged electron on the chlorine atom makes that atom a negative chloride ion, and the loss of an electron by the sodium atom leaves a net positive charge on that ion. The question "Why?" cannot be answered so easily. The answer invokes much of the present physical explanation of chemical behavior and will emerge in the next few chapters.

Many other crystals—especially of materials that the chemist calls "salts"—are held together by similar ionic bonds and are called *ionic crystals*. But clearly ionic bonds can be found only in crystals containing atoms of at least two different elements that can form the ions of opposite charge. Other explanations must be found for crystals made of a single element, such as diamond, copper, or the solids that the rare gases such as neon form at low temperatures.

Van der Waals Forces

Solid neon may seem so odd and out of the way as hardly to be worth attention. But it offers an interesting problem with an interesting answer. Here surely are some of the most nearly inert atoms that we know. Nevertheless, at low temperatures the atoms cohere into a solid—fragile and low-melting to be sure but bonded to some extent in some fashion. It turns out that solid neon is bonded by a force between its atoms that is almost always present in other solids as well—a force much stronger than gravitation but weak enough to be masked wherever other bonding forces, such as the ionic force, are also at work.

Visualize a neon atom as a positively charged nucleus surrounded by a spherical cloud of negative charge formed by the electrons in the atom (Fig. 2). On the average (Fig. 2a), the nucleus is at the center of that spherical cloud. But the electrons are moving about, and instantaneously (Fig. 2b) the cloud of negative charge may not be centered exactly at the nucleus. Whenever the center of

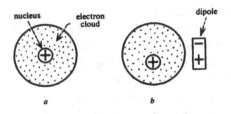

FIG. 2—A cloud of electrons, moving about the nucleus of an atom, (a) centered on the average but (b) instantaneously uncentered.

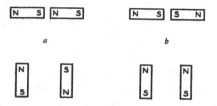

FIG. 3 – Two bar magnets (a) attracting each other in some arrangements and (b) repelling in others.

the charge cloud does not coincide exactly with the nucleus, the atom is an *electric dipole.*

Perhaps you are more familiar with magnetic than with electric dipoles; most people have some acquaintance with the behavior of bar magnets. You can liken the interaction of two atoms, each with an electric dipole, to the interaction of two little bar magnets. The magnets will repel each other if they are arranged as in Fig. 3b and attract each other if they are arranged as in Fig. 3a; in other arrangements they will show intermediate behavior.

The forces between two electric dipoles can be analyzed in somewhat the same way as the forces between the ions in a sodium chloride crystal were analyzed. Each dipole consists of two charges equal in magnitude and opposite in sign. The force between the two dipoles will be the net result of the forces exerted by each of the two charges in one dipole on each of the two charges in the other. Figure 4 shows how to examine two arrangements of the two dipoles, which are especially easy to understand.

In Fig. 4a the positive charge in the dipole at left (above) is slightly nearer to the positive than to the negative charge in the dipole at the right, and hence that charge repels the dipole slightly. Similarly, the negative charge in the dipole at the left also repels the dipole at the right, and thus the two dipoles repel each other. In Fig. 4b, the attractions between the charges slightly outweigh the repulsions, and the two dipoles attract each other.

The electric dipoles belonging to any pair of neon atoms are fluctuating in time, of course, but on the average they will attract each other more of the time than they will repel each other because their energy of interaction is lower when they attract. All atoms,

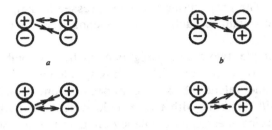

FIG. 4 — (a) A repelling arrangement and (b) an attractive arrangement of two dipoles.

ions, and molecules have such fluctuating dipoles; and when they come close to one another, their dipoles couple together in a way that provides a little attractive force bonding the atoms to one another. For neon, calculation shows that this force alone is adequate to account for the cohesion of its atoms when it solidifies.

The bonds from this attractive force between atoms are often called *van der Waals bonds*. The name honors the Dutch physical chemist who proposed a modification of the perfect gas law mentioned in Chapter II — the relationship connecting the pressure, volume, and temperature of a gas. Van der Waals' proposal takes account of the fact that when gases are not dilute, their behavior deviates from the perfect gas law, partly because attractive forces of this sort become noticeable.

But a force so small is quite inadequate to explain the rigidity and tenacity of the interatomic bonding in diamond or in the metals. There the forces are often as large as those holding the atoms together in the molecules of chemical compounds. It is natural to suppose that such strong bonding in solids is closely allied with the bonding of atoms into molecules.

Valency

In following the path suggested by this alliance, we encounter first the observations of atomic behavior that have been correlated as rules of *valency*. The idea of valency was originally derived from chemical analyses of compounds. Quantitative analyses show that the various species of atoms, the chemical elements, usually combine with one another in fixed relative proportions as Chapter II described. The proportions can be expressed in simple fractions if the atomic weight of each element is chosen as the unit of weight for it.

For example, two atomic weights of hydrogen combine with one of oxygen to form water. Accordingly valency 1 is ascribed to hydrogen and valency 2 to oxygen, because one atomic weight of oxygen will combine with two atomic weights of hydrogen. Any atomic species is assigned a valency equal to the number of atomic weights of hydrogen that will combine with one atomic weight of the species.

Correspondingly, the atomistic picture of water portrays it as a collection of water molecules, each containing two atoms of hydrogen and one of oxygen. Often the water molecule is diagrammed as H—O—H, with a bond connecting each atom of hydrogen to the atom of oxygen. Then the valency of each atom is given by the number of bonds in which it participates.

Carbon then receives valency 4, in recognition of the fact that one atomic weight of carbon combines with four atomic weights of hydrogen to form methane:

This assignment makes the valency of carbon twice that of oxygen. The requirement that these assignments be consistent is satisfied in the fact that carbon and oxygen combine to form carbon dioxide, diagrammed as O=C=O. To be sure, carbon and oxygen will also combine in a one-to-one proportion into carbon monoxide. But carbon monoxide behaves in many ways as if it were somewhat unsaturated; for example, it readily burns in oxygen to form carbon dioxide.

In order to make the number of bonds formed by each atom equal to the valency of its species, the bonding diagram of a carbon dioxide molecule must connect the carbon atom with *double bonds* to each oxygen atom. Such a diagram immediately provokes the question, Is a double bond twice as strong as a single bond? The answer comes from comparing the energy required to break a double bond in some kind of molecules with the energy required to break a single bond between atoms of the same elements in some other kind of molecules. The comparison shows that a double bond is often almost twice as strong as a single bond. In other words, a method of diagramming, originated merely to display the relative combining weights of atomic elements, turns out to represent also some features of the strengths of the interatomic forces.

Proceeding through the list of atomic elements, we can set up a consistent scheme of valencies that describes remarkably well the chemical behavior of many of the elements, especially those of lowest atomic weight. Models of chemical molecules are often built out of rods stuck into holes in balls that represent the atoms. Each rod represents a bond, and the number of holes in each ball is made equal to the valency of the atom that it represents.

Useful as these rods are in providing a consistent formal device for bringing the facts of chemistry into order, they still do not give a physical explanation of the bonds. They are strongly reminiscent of the "hooked Atoms" castigated by Newton as "begging the Question." To be sure, since a bond represented by two rods turns out in many cases to be nearly twice as strong as a bond represented by one rod, the rods represent the bonding surprisingly well. But it was only at the beginning of this century that any deeper physical insight into chemical bonds began to emerge, as a byproduct of the discovery of how atoms are built out of nuclei and electrons.

The Periodic Table

Studies of valency had already shown that the different elements can be classified into groups. The elements within each group have the same valency, form similar chemical compounds, and behave similarly in chemical reactions. Especially at the hands of Dmitri Mendeleyev and Lothar Meyer, the elements were tabulated in a way that correlated the grouping with their atomic weights. In that table the elements are listed in order of increasing atomic weight from left to right, like ordinary reading matter, and a new line is begun when the valency begins to repeat. The elements falling under one another in the vertical columns form the groups.

This procedure is especially successful for the lighter elements.

At the left side of Table 2 are the elements of valency 1, and at its extreme right are the chemically inert rare gases, such as neon, of valency 0. The elements on the left tend to combine with those on the right to form salts, which solidify in ionic crystals. The positive or negative sign attached to each valency in Table 2 refers to the electrically positive or negative character of the ions formed by the atoms with that valency.

As the listing was continued to include the heavier elements, it was found necessary to interrupt the simple periodicity of the table by inserting *transition periods* that accommodate successions of elements that increase in atomic weight but have similar chemical properties. You may find it helpful to look first at the abbreviated Table 3, which omits the transition series. The chemical similarity of the elements in any one group becomes conspicuous. The alkali metals in Group I all form strongly basic hydroxides closely resembling sodium hydroxide, the substance popularly called "lye." The species in Group VII are called *halogens* because they will combine with the alkali metals to form salts resembling *halite*, the mineralogist's name for common salt.

The principal transition series interrupt the abbreviated table between Groups II and III in its fourth, fifth, and sixth periods. Chemists often assign the elements in these transition series to.the eight major groups (Table 4). The assignments recognize that these elements frequently show principally the valencies appropriate to these groups. Thus, copper (Cu), silver (Ag), and gold (Au) often show valency 1.

But copper more often shows valency 2, silver can also show that valency, and gold even shows valency 3. Indeed, such ambivalence characterizes most of the transition species. The currency metals—copper, silver, and gold—resemble one another more than they resemble the alkali metals, and in recognition of such relationships each major group in the Periodic Table is usually divided into two subgroups, as shown in Table 5.

The assignment of the three similar elements, iron, cobalt, and nickel, to Group VIII is a final admission of failure to force Nature into so simple a mold, for these metals bear no chemical resemblance whatever to the rare gases. Finally, the last of the principal trasition periods is itself interrupted by the 14 rare earth elements,

TABLE 2

Group:	I	II	III	IV	V	VI	VII	VIII
Valency:	+1	+2	+3	4	−3	−2	−1	0
Period 1	1 H	...	...	...	...	...	...	2 He
Period 2	2 Li	4 Be	5 B	6 C	7 N	8 O	9 F	10 Ne
Period 3	11 Na	12 Mg	13 Al	14 Si	15 P	16 S	17 Cl	18 A

No.	Name	Symbol	Weight	No.	Name	Symbol	Weight
1	Hydrogen	H	1.0	10	Neon	Ne	20.2
2	Helium	He	4.0	11	Sodium	Na	23.0
3	Lithium	Li	6.9	12	Magnesium	Mg	24.3
4	Beryllium	Be	9.0	13	Aluminum	Al	27.0
5	Boron	B	10.8	14	Silicon	Si	28.0
6	Carbon	C	12.0	15	Phosphorus	P	31.0
7	Nitrogen	N	14.0	16	Sulfur	S	32.1
8	Oxygen	O	16.0	17	Chlorine	Cl	35.5
9	Fluorine	F	19.0	18	Argon	A	39.9

TABLE 3

Group:	I	II	III	IV	V	VI	VII	VIII
Valency:	+1	+2	+3	+4 −4	(+5) −3	(+6) −2	(+7) −1	0
Period 1	(1 H)	...	...	...	...	...	1 H	2 He
Period 2	3 Li	4 Be	5 B	6 C	7 N	8 O	9 F	10 Ne
Period 3	11 Na	12 Mg	13 Al	14 Si	15 P	16 S	17 Cl	18 A
Period 4	19 K	20 Ca	31 Ga	32 Ge	33 As	34 Se	35 Br	36 Kr
Period 5	37 Rb	38 Sr	49 In	50 Sn	51 Sb	52 Te	53 I	54 Xe
Period 6	55 Cs	56 Ba	81 Tl	82 Pb	83 Bi	84 Po	85 At	86 Rn

TABLE 4

Group:	III	IV	V	VI	VII	VIII	I	II
Period 4	21 Sc	22 Ti	23 V	24 Cr	25 Mn	26 Fe 27 Co 28 Ni	29 Cu	30 Zn
Period 5	39 Y	40 Zr	41 Nb	42 Mo	43 Ma	44 Ru 45 Rh 46 Pd	47 Ag	48 Cd
Period 6	57 La	72 Hf	73 Ta	74 W	75 Re	76 Os 77 Ir 78 Pt	79 Au	80 Hg

TABLE 5

Group:	I	II	III	IV	V	VI	VII	VIII
Period 1	(1 H)	...	...	...	...	...	1 H	2 He
Period 2	3 Li	4 Be	5 B	6 C	7 N	8 O	9 F	10 Ne
Period 3	11 Na	12 Mg	13 Al	14 Si	15 P	16 S	17 Cl	18 A
Period 4	19 K	20 Ca	21 Sc	22 Ti	23 V	24 Cr	25 Mn	26 Fe 27 Co
	29 Cu	30 Zn	31 Ga	32 Ge	33 As	34 Se	35 Br	28 Ni 36 Kr
Period 5	37 Rb	38 Sr	39 Y	40 Zr	41 Nb	42 Mo	43 Ma	44 Ru 45 Rh
	47 Ag	48 Cd	49 In	50 Sn	51 Sb	52 Te	53 I	46 Pd 54 Xe
Period 6	55 Cs	56 Ba	57 La	72 Hf	73 Ta	74 W	75 Re	76 Os 77 Ir
	79 Au	80 Hg	81 Tl	82 Pb	83 Bi	84 Po	85 At	78 Pt 86 Rn

whose chemical resemblance to one another is so close that any one of them can be separated from the others only with difficulty.

Planetary Electrons

The beginnings of a physical explanation for this behavior of the elements — for their valencies and for the periodicity of their properties with increasing atomic weight — appeared in the early years of this century. A helpful picture of atoms was proposed, particularly by Sir Ernest Rutherford and Niels Bohr, to set in order the results of spectroscopic studies far removed from chemistry.

In that picture an atom is made of a nucleus that carries most of the atom's mass and relatively light electrons circulating around the nucleus. Each nucleus bears a positive electrical charge and each electron a negative charge. All electrons, wherever they are found, are alike, in contrast to the nuclei, which have different masses and different amounts of positive charge in the different elements. Under most circumstances, an atom has no net charge, because the negative charges on its electrons exactly balance the positive charge on its nucleus. The electrons are held around the nucleus by the electrostatic attraction between nucleus and electrons.

The earliest forms of this picture portrayed the electrons as tiny particles traversing well-defined orbits, as do the planets traversing orbits about the sun in the solar system. Indeed, there should be many close correspondences between the behavior of two particles such as a nucleus and an electron, attracting each other by an electrostatic force, and the behavior of a pair of objects, such as the sun and a planet, attracting each other by a gravitational force. These two sorts of attractions share one important property; they vary in the same way with the distance between the attracting objects. Both obey the inverse-square law of force, already mentioned in reference to the ionic bond. Hence, the shape of an electronic orbit, for example, would be like that of a planetary orbit.

In such a picture the only differences between atoms and planetary systems, apart from their very different sizes, would arise from the fact that the planets are attracted by one another, whereas the electrons are repelled by one another. The planets disturb one another's orbits by pulling one another; the electrons would disturb the orbits by pushing one another.

In fact, however, studies of the wavelengths of light emitted by glowing matter showed that there is another difference — a much more important difference — between an atom and the solar system.

In order to explain the observed spectra of such light, Bohr had to assume that the shapes and sizes of the orbits which electrons can traverse in an atom are much more strictly limited than the orbital shapes permitted to a planet.

This limitation of an electron in an atom is a *quantum restriction.* It is somewhat like the limitation that permits a harmonic oscillator to vibrate only at certain amplitudes, a limitation that was used in Chapter III to explain the low heat capacity of solids at low temperatures. The physical origin of these quantum restrictions was obscure when they were first suggested, but they have since become a little better understood, as Chapter XI will describe.

Unfortunately, recent work has removed some of the definiteness that is so attractive in the picture of planetary electrons. An electron cannot be localized as precisely as the picture of an orbit suggests; it must be smeared somewhat over the space that it inhabits. But the results, derived from the earlier picture of definite orbits strictly limited in their permitted shapes, still provide a useful way to correlate many observations of atomic behavior.

Shells of Orbits

In particular, that picture classified the permitted orbits into sets. Each set was described as occupying a shell in the space around the nucleus of the atom; in the more recent theories an electron in an atom can still be assigned to one or another of those shells. Each shell has a limited capacity for electrons; the innermost shell can hold no more than two electrons, for example, and the next shell no more than eight (Fig. 5).

The succession of atomic species in the Periodic Table thus represents the successive filling of these shells by electrons. A

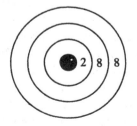

FIG. 5 – Quotas of the first three shells of electronic orbits in an atom.

hydrogen atom has only one planetary electron circulating around the *proton* that forms its nucleus. A helium atom has two electrons, circulating around a nucleus with twice the charge and four times the mass of the proton. Two electrons complete the quota of the innermost shell of orbits, and thus the additional electron in the lithium atom circulates in one of the orbits in the next shell. The maximum quota of that shell is completed in neon; therefore, occupancy of still another shell of orbits begins in sodium. Figure 6 portrays the successive steps of shell-filling in the first three periods of the Periodic Table.

As this picture of atoms developed, a way was found for using it to explain the chemical rules of valency. That way focused attention on the relatively great inertness of the rare gases toward forming chemical compounds. If an atom has especially great stability when its electronic shells have exactly their maximum quota of electrons, then it may tend to gain or lose electrons in order to achieve that stability. Thus, chlorine, lacking one electron to fill its outermost occupied shell, might tend to pick up that electron. Sodium, with only one electron in its outermost occupied shell, might readily lose that electron in order to leave the shell empty and expose the filled shell beneath as its outermost occupied shell. This picture (Fig. 7) easily explained the fact that sodium chloride is made of ions, not of neutral atoms.

The explanation of non-ionic combinations of atoms required a further step. It was assumed that two atoms, both with outer shells only partly filled, could also fill those shells by *sharing* electrons

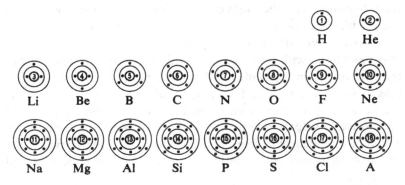

FIG. 6—Shell-filling in the first three periods of the Periodic Table.

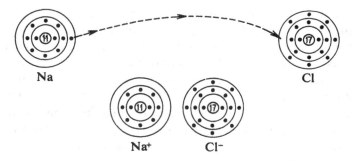

FIG. 7 – Shell-filling in sodium chloride.

with each other. Hydrogen gas, for example, is not made of independent atoms of hydrogen but of molecules of hydrogen, each containing two atoms tightly bonded together. By sharing electrons, each hydrogen atom provides its innermost shell with two electrons instead of one, and thus that shell is filled like the corresponding shell of the rare gas helium.

At first this explanation may seem to require too many electrons. If each of the two atoms in a hydrogen molecule has two electrons in a shell, there may seem to be four electrons, and the two atoms have only two electrons to contribute. But, of course, there are still only two electrons. A shell contains permitted orbital possibilities; it is a shell of orbits, not necessarily a shell of electrons. Each hydrogen atom offers two orbital possibilities in its innermost shell, and there are two electrons available. Either electron can move from its orbit in one of the atoms, spend a little time circulating in the other atom, and then move back. This is the sense in which the shell is filled by the native electron and the visiting electron.

You will have little difficulty accepting the notion that atoms can share electrons with one another in this way when you remember that the atoms in molecules and solids are very close to one another. Since every atom has a positively charged nucleus at its center, an electron can be expected to stray occasionally to any of the attractive neighboring nuclei. The next chapter describes when and how the electrons stray and how their straying provides the electrostatic forces that hold the atoms together in molecules and in metals.

VII. MOLECULES AND METALS

In a metal, however, an atom is not greatly restricted
in regard to the number and direction of the bonds
which it can form with adjacent atoms.
LINUS PAULING, *The Nature of the Chemical Bond*

WHY DISCARD the idea that the hydrogen molecule is held together by an ionic bond? Such a picture would violate the sense of symmetry. It seems unsuitable to suppose that two hydrogen atoms in every way alike should engage in partnership by becoming unlike.

To be sure, the symmetry in a molecule of hydrogen is an average symmetry, not a symmetry maintained at every instant. The electron on one of the hydrogen atoms may stray to the other one once in a while and give it a negative charge instantaneously. But just as often that situation will be reversed; on the average the molecule will be symmetrical. The molecule's appearance is somewhat similar to that of the neon atom, described in the last chapter, which is a dipole instantaneously, spherically symmetrical on the average. But the molecule must be likened to a dumbbell instead of a ball. Indeed, whenever two like atoms share electrons to form a diatomic molecule, the two atoms remain alike and hold some of their electrons in common.

Sharing Electrons

It is not true, however, that atoms of any element will share electrons with the atoms of any other element or even with its own kind. As the end of the last chapter described, they can be expected to share when they would thereby attain filled shells. This idea can be most simply applied to the lighter chemical elements — those in the first part of the Periodic Table.

Each of the two rows of the Periodic Table following helium contains eight elements. Much of the chemical behavior of these elements can be summarized in the assertion that they will combine with one another so that each atom acquires eight electrons in its outermost occupied shell either by capturing electrons, by losing electrons, or by sharing electrons.

For example, a neutral chlorine atom has seven electrons in a shell that can hold eight; a neutral sodium atom has in its outermost occupied shell one electron, whose loss would put outermost the underlying shell occupied by eight electrons. Hence, the sodium atom can contribute an electron to the chlorine atom to form the ionic sodium chloride bond (Fig. 7 of the last chapter). In the diatomic molecules that form chlorine gas, on the other hand, each chlorine atom shares one of the seven electrons in its outermost occupied shell with its partner in the molecule, so that the outermost shell of each atom acquires its full complement of eight electrons.

A convenient way to show how the electron-sharing idea can be applied in any instance is to write the conventional symbol for each of the atoms in question, place dots around the symbol to stand for

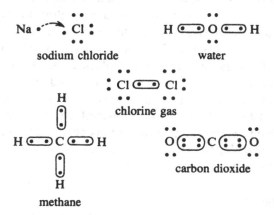

FIG. 1 — Closed shells of electrons in atoms forming molecules.

the electrons in its outermost occupied shell, and hook two dots together to show how a pair of electrons is shared or draw an arrow to show how an electron is transferred to form two ions. Figure 1 uses these conventions to formalize the behavior of the atoms in sodium chloride, water, chlorine gas, methane, and carbon dioxide.

An interesting result of the electron-sharing idea is that it can explain the existence of certain chemical compounds that do not obey the simpler rules of valency. Ammonia and boron trifluoride are both familiar chemical compounds, in which nitrogen and boron display their expected valencies of 3:

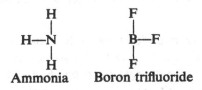

Ammonia   Boron trifluoride

But a molecule of ammonia will join a molecule of boron trifluoride to form a molecule of a new compound, in which both nitrogen and boron display the unexpected valency of 4.

When the ammonia and boron trifluoride molecules are written to show how the electron-sharing in each of them might be arranged (Fig. 2), it becomes clear how an additional bond might arise. There

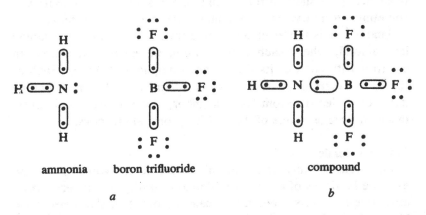

FIG. 2 — (a) Boron trifluoride and ammonia forming (b) a donor-acceptor bond.

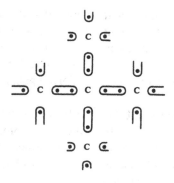

FIG. 3 – Electron-sharing in diamond.

is a deficiency of two electrons in the outermost occupied shell of the boron atom in a molecule of boron trifluoride. There are two electrons that are not engaged in bonding in the outermost occupied shell of the nitrogen atom in a molecule of ammonia. By allowing its nonbonding pair of electrons to be shared by the boron atom, the ammonia molecule forms a bond to the boron trifluoride molecule.

A bond of this sort is sometimes called a *donor-acceptor* bond; nitrogen is the donor, boron the acceptor, of a pair of shared electrons. Since an atom of oxygen has only six electrons in its outermost occupied shell, nitrogen compounds similar to ammonia will sometimes form oxides by bonding to oxygen in this fashion.

Among solids the electron-sharing idea finds its purest example in diamond, where each carbon atom, of valency 4, shares an electron with each of its four nearest neighbors (Fig. 3). Each of the four electrons in its outermost occupied shell serves double duty. One electron from each neighbor, also serving double duty, takes advantage of one of its additional orbital offerings.

Covalent Bonds

The great stability of a diatomic molecule of hydrogen and the extreme hardness of a crystal of diamond testify that a very strong attractive force arises between a pair of atoms that share electrons. You can quickly see the physical nature of that force when you look more closely at how an electron is·shared.

In the hydrogen molecule, for example, the electron on one of the atoms strays to the other atom occasionally, finds an unoccupied orbit in the innermost shell of the atom that it visits, and circulates in that atom for a short time. While it is doing this, it puts its negative charge on its host and leaves a net positive charge on the atom that it has abandoned. Then the two atoms attract each other just as two ions would. The other electron can behave in the same way; in that case the charges on the two atoms are reversed but again provide an attractive force.

If both electrons were to stray at the same time, simply exchanging places, they would put no net charge on either atom. But in passing from one atom to the other, they must spend a little time between the two. During that little time, both atoms would have a net positive charge, and both would be attracted toward each other by the negative charge on the two electrons dashing from one to the other. Thus, however the two atoms accomplish the sharing of electrons, there is an electrostatic attractive force between them (Fig. 4).

You may wonder what fraction of the attractive force comes from ionic arrangements like those of Fig. 4a and b and what fraction from arrangements like that of Fig. 4c, in which the electrons are between the atoms. There has been much disagreement about the answer, and no doubt the right answer varies from one element

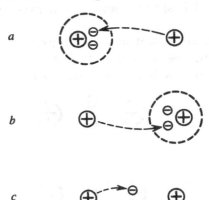

FIG. 4—Three formal arrangements of the two electrons in a hydrogen molecule.

to another. But for hydrogen there is a bit of evidence strongly suggesting that arrangements like Fig. 4c — called *covalent* arrangements — are the more influential.

The One-Electron Bond

That evidence comes from the *hydrogen molecule-ion*. The hydrogen molecule-ion is a hydrogen molecule from which one electron has been removed, leaving it with a net positive charge. Thus, it is the simplest imaginable molecule; it is composed of two nuclei and one electron. When its one electron is on either of the atoms, that atom is electrically neutral and does not attract the other. When the electron is between the atoms, it attracts both but attracts them only half as strongly as two electrons would (Fig. 5). And, in fact, the measured energy required to break apart the hydrogen molecule-ion is about half the energy required to break the hydrogen molecule.

You must be careful not to take these conventional diagrams too literally. The electron is never resting comfortably either on one atom or midway between two. It is coursing some complicated orbit, which takes it about both atoms so rapidly that the much heavier nuclei cannot respond to the variations in the direction of the electron's attraction. The nuclei can respond only to the average force that the electron exerts on them. And the electron will bond the nuclei if it spends more time between them than elsewhere (Fig. 6).

Bonds between atoms that share electrons are called *covalent*, or *homopolar*, bonds. The bond in the hydrogen molecule-ion belongs to a subspecies of covalent bond called a *one-electron* bond to

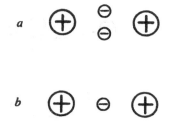

FIG. 5 — (*a*) The hydrogen molecule. (*b*) The hydrogen molecule-ion.

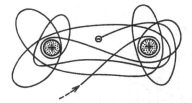

FIG. 6—Imagined motion of electron bonding a molecule.

distinguish it from the *electron-pair* bond in the hydrogen molecule and in the molecules of almost all other chemical compounds.

The prevalence of electron-pair bonds has often been rashly interpreted as betokening a mysterious force that comes into being with the pairing of electrons. But the hydrogen molecule-ion dispels that mystery by showing that a bond approximately half as strong as an electron-pair bond arises from half as many electrons. The reason that electron-pair bonds are more common is simply that there are usually enough electrons available to form them whenever atoms offer suitable unoccupied orbits.

Metallic Cohesion

Studying the hydrogen molecule-ion is helpful in still another way; it provides preparation for understanding how the atoms are bonded in solid metals. In the hydrogen molecule-ion, each of two atoms offers two orbital possibilities in its bonding shell. If the two atoms shared two electrons, the orbital possibilities would be completely utilized, but there is only one electron available. In solid metals the atoms provide an even larger excess in the number of orbital possibilities over the number of electrons to fill them.

Sodium, for example, forms metallic crystals with the structure shown in Fig. 7. Each sodium atom has eight others as immediate neighbors. Since the outermost occupied shell of a sodium atom contains only one electron, the arrays of neighbors offer many more possible orbits than electrons to fill them. For this reason the bonds in the hydrogen molecule-ion and in the metals are often called *electron-deficient bonds*.

A little arithmetic may help to make definite the idea of the electron-deficient bond. First, suppose that each atom attempts to

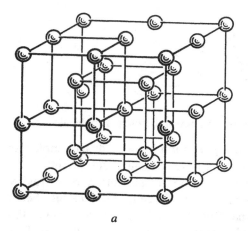
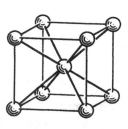

a   b

FIG. 7—Atoms in a crystal of sodium.

make a bond with each of its immediate neighbors. Then count the number of bonds per atom that the structure must have. Finally, count the number of electrons available per bond. If that number is less than 2, then the electron deficiency in the bonding is 2 minus that number, and that number itself is the *electron occupancy* in the bonding.

For instance, in diamond each carbon atom has 4 neighbors. Each bond involves a pair of neighbors, and therefore there are 2 bonds per atom. Each atom has 4 electrons in its outermost occupied shell. Since there are 4 available electrons per atom, and 2 bonds per atom, there are 2 available electrons per bond—just the right number for *saturated bonding* with no electron deficiency. In the crystal of metallic sodium, on the other hand, there are 8 neighbors and one available electron per atom. Thus, there are 4 bonds per atom and only $\frac{1}{4}$ electron per bond; the electron deficiency is $1\frac{3}{4}$ per bond.

The fact that the electrons in a metal can move freely from atom to atom explains the high electrical conductivity of metals. Metals are *electronic conductors* that conduct electricity by the motion of free electrons and thus are unlike a salt solution, which conducts electricity by the motion of free ions. In solid salt the ions cannot move freely because they are much bulkier than electrons and are packed together tightly, and the crystal is an insulator. But when salt is melted, the ions can move past one another, and the molten material becomes a good ionic conductor.

The freedom of the electrons to move in a metal makes them behave in some ways as do the atoms in a gas, dashing about in their container. But there are two conspicuous differences between an electron gas and an atomic gas. In the first place, the electron gas is compressed into a much smaller space than the atomic gas. In sodium, with one free electron per atom, there are as many electrons as atoms in a given amount of space. The electron gas is compressed as much as an atomic gas that has been squeezed down to 10^{-3} of its normal volume.

In the second place, unlike an atom in a gas, each electron bears a negative charge. The electrons are all repelling one another and trying to stay out of one another's way. And they have left behind them their parent atoms, which have thus become positive ions. But the electrostatic attraction between those ions and the electron gas outweighs the mutual repulsion of the electrons and thus holds the entire assembly together.

A third difference between an electron gas and an atomic gas is even more important than these two, but its explanation must await the discussion of the modern theory of electronic behavior in a later chapter.

Mixed Bonding

So far, the last two chapters, describing four extreme types of bonding, have shown that all can be analyzed into electrostatic attractions between opposite electric charges. Sometimes it is helpful to have a simple way of visualizing the four conventional extremes. With the warning that it must be understood in terms of the preceding discussions, Fig. 8 is an aid for recalling the meaning of the *ionic* bond, the *van der Waals* bond, the *covalent* bond, and the *metallic* bond.

There are a few instances in which each of these types of bonds occurs in almost pure form, but usually a bond comprises a mixture of several types. For example, as the discussion of van der Waals forces emphasized, nearly every interatomic attraction has a van der Waals ingredient. Every atom except hydrogen has an innermost shell of orbits completely occupied by two electrons. Pro-

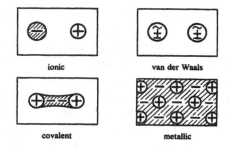

FIG. 8—The four extreme types of bonding between atoms.

ceeding down the Periodic Table, you find that more and more shells become filled. In other words, all of these atoms have groups of electrons in inner shells, which do not engage in ionic, covalent, or metallic bonding. But those buried electrons do produce fluctuating dipole moments that couple with the similar dipole moments of neighboring atoms. Only hydrogen atoms, with no inner electrons, are exempt from engaging in van der Waals attractions with the atoms to which they are otherwise bonded in a molecule.

Even hydrogen loses that exemption when two hydrogen atoms are joined in a hydrogen molecule. That molecule, like a helium atom, has two electrons bound to two positive charges. The fluctuating dipole moments of hydrogen molecules provide attractive forces that hold the molecules together at very low temperatures in liquid and solid hydrogen.

More important than the admixture of van der Waals attraction, however, is the intermixing of ionic, covalent, and metallic ingredients in strong bonds. In a covalent bond between atoms of two different elements, for example, the sharing of a pair of electrons will usually be unequally distributed. The shared pair may spend more time near one atom than the other, giving the covalent bond an ionic ingredient.

By looking back at the Periodic Table, you can see one sort of situation in which a mixed ionic-covalent bond will appear. Sodium and chlorine, from the extreme left and right of the table, bond ionically to each other. Carbon, from the middle of the table, bonds covalently to itself. Atoms like magnesium and sulfur, partway to the left and partway to the right, will form a mixed bond, in which the magnesium atoms do not wholly lose their two outermost electrons to the sulfur atoms and the ionic bond has a large covalent ingredient.

There is a simple electrostatic explanation for the transition from ionic to covalent bonding between atoms of elements that are nearer the center of the Periodic Table. In picking up one electron, a sulfur atom acquires a negative charge which repels a second electron. In losing that second electron, a magnesium ion doubles its positive charge and thus doubles the attractive force pulling the second electron back. Hence, ions whose formal charges are greater than one electronic unit seldom behave like true spherical ions possessing their full expected electrical charge. The electrons stray from the negative ions to the positive ions, reducing the effective charges of both.

Polarization Forces

Even when a positive ion cannot recapture an electron from a negative ion, the attractive force on the cloud of electrons in the negative ion is noticeable. That force distorts the orbits that the electrons would traverse in the negative ion if it were isolated and shifts its electron cloud a little toward the positive ion. Thus, the negative ion acquires a dipole ingredient in addition to its charge. The induced dipole, unlike the fluctuating dipoles that provide the van der Waals attraction, is fixed in direction along the line of the two ions. The positive ion is said to have *polarized* the negative ion.

A negative ion performs an analogous act on a positive ion; the negative charge repels the cloud of electrons on the positive ion and polarizes it. Figure 9 shows the direction of the dipoles induced in both ions. Comparing Fig. 9 with Fig. 3a of the last chapter, you will notice that the force between the induced dipoles augments the attraction between the two ions.

Among positive ions, hydrogen again takes a unique position. When a hydrogen atom loses its electron to some atom that forms a negative ion, the remaining hydrogen ion consists simply of a proton. Since it has no electronic cloud, the hydrogen ion cannot be polarized by the negative ion. But it can polarize the negative ion and so add, to the ionic attraction, the *ion-dipole force* analyzed in Fig. 10a.

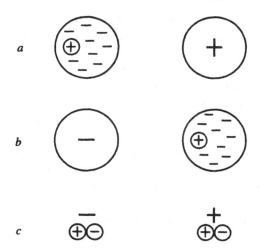

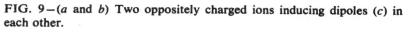

FIG. 9—(*a* and *b*) Two oppositely charged ions inducing dipoles (*c*) in each other.

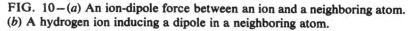

a   *b*

FIG. 10—(*a*) An ion-dipole force between an ion and a neighboring atom. (*b*) A hydrogen ion inducing a dipole in a neighboring atom.

Of course, hydrogen is not unique in producing an ion-dipole force; whenever any ion polarizes another, an ion-dipole force arises. But the force can be especially great when a hydrogen ion participates. A proton is much smaller than all other ions because those others retain electrons. The proton can put its positive charge very much closer to a negative ion and polarize it much more effectively (Fig. 10*b*). The resulting distortion of the electronic cloud of the negative ion is hardly distinguishable from the sharing of electrons between a hydrogen atom and another atom.

Positive ions can polarize not only negative ions but also neutral atoms, and again protons are especially effective. A proton is so small that it can bring two other atoms close together by getting between them, polarizing them, and attracting them by way of the dipoles that it has induced in them. Two such atoms are said to be *hydrogen bonded*. Oxygen atoms especially are attracted toward each other by hydrogen bonds; no doubt, crystals of ice and of sugar, for example, are held together mainly in that way.

In fact, the study of crystalline solids provides many of the best examples of the types of bonding that this chapter has discussed. The character of the bonds between their atoms has a strong influence on the character of the orderliness that the atoms adopt, as the next chapter will show.

VIII. STRUCTURES

*Since every solid substance contains parts that are crystalline,
and since in many of them the whole is an aggregation of
crystals, it will be readily understood that a knowledge of
crystal structure often affords an explanation of
the properties of the substance.*
SIR WILLIAM BRAGG, *The Universe of Light*

IT IS EASY to argue in the following fashion. When a solid is made
of a single kind of atom, the fact that it is solid shows that the
atoms are all attracting one another. Then each may be expected to
collect around it as many of its fellows as it can. If the atoms are
shaped like spheres, their arrangement in the solid should be one
that identical spheres will take when they are packed as closely as
possible.

In 1665 Robert Hooke showed in his book *Micrographia* a
comparison (Fig. 1) of the shapes of alum crystals with the shapes
taken by close-packed spheres. Extending the comparison into
three dimensions, he pointed out that the regular tetrahedron often
appearing on an alum crystal might be connected with an arrangement of one ball resting on three others, all in contact.

Such reasoning cannot yield the arrangement of atoms in so
complicated a material as alum; in fact, a unit cell of that substance
has been found to contain no less than 192 atoms. In general the
determination of a crystal structure by X rays will verify only a
prediction that balances many factors, principally the sizes and
shapes of the molecules and the types of bonding between them.
But if Hooke had ascribed his guess to metallic copper, he would
have been right.

Close-Packed Spheres

In crystals made of atoms of a single element the atoms often adopt close-packed arrangements resembling those Hooke pictured. You can expect to find those structures wherever the forces between the atoms have no directional limitations, and the atoms all attract their fellows indiscriminately. For example, the van der Waals attractions described in Chapter VI, between the atoms of the rare gases such as neon and argon, bring those atoms together into close-packed structures at low temperatures.

Sometimes a whole molecule will behave like a spherical atom. The molecules of methane are made of carbon atoms, each covalently bonded to four hydrogen atoms (Fig. 1 of the last chapter). Those molecules are nearly spherical and form crystals with a close-packed structure, bonded by the van der Waals attractions between the molecules rather than the individual atoms.

Many common metals, such as magnesium and copper, also crystallize in close-packed structures. The negatively charged electron gas often brings the positively charged metal ions as close together as possible. But although they are both close-packed, the

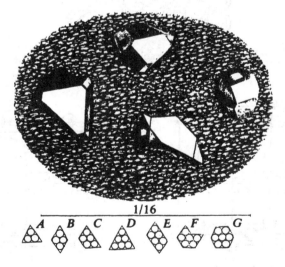

FIG. 1—Hooke's comparison of alum crystals with piles of close-packed spheres.

atomic arrangements in magnesium and in copper differ; there is more than one orderly way of packing spheres closely.

To see how that can be, look first at close-packing in two dimensions and then at how successive close-packed layers can be built one upon another. In two dimensions, spheres pack as Fig. 2 shows; each sphere touches six others. When such a layer of spheres is covered by another layer, each sphere of the second layer settles into a triangular depression formed by three spheres in the first, and thus each sphere acquires three more neighbors. When another layer is placed on the opposite side of the first, the spheres in the first are completely embedded, and each finds itself in contact with 12 immediate neighbors.

But the third layer can be added in two ways. In one way (Fig. 3a) spheres in the two outer layers of the sandwich fall directly opposite one another; in the other way (Fig. 3b) they do not. Con-

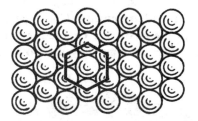

FIG. 2 — A close-packed layer of spheres.

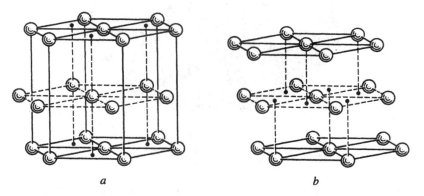

a   b

FIG. 3 — (a) Hexagonal close packing. (b) Cubic close packing.

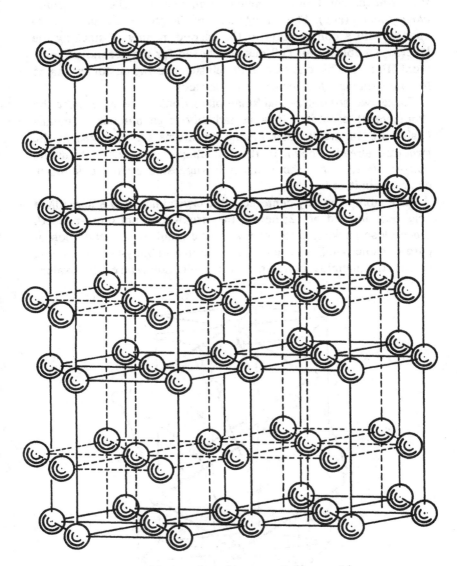

FIG. 4—Continuation of hexagonal close packing.

tinuation of the first manner of stacking the layers (Fig. 4) produces the arrangement found in magnesium, usually called *hexagonal close-packed* (hcp). The arrangement of the atoms in copper corresponds with a continuation of the second manner of stacking the layers. A cubical unit cell can be chosen for the second arrangement (Fig. 5). This choice of unit cell suggests the customary name for the structure, *face-centered cubic* (fcc).

These two structures and small distortions of them describe the arrangement of atoms in a remarkably large number of metals (Table 6). In many other metals the atoms are arranged in the structure already shown in the last chapter as Fig. 7. Its name, *body-centered cubic* (bcc), again is suggested by the customary choice of unit cell.

Each atom in a body-centered cubic crystal has only 8 immediate neighbors instead of 12; hence, the structure is not close-packed. Since close packing is the characteristic result of the operation of pure metallic bonding, you can assume that in body-centered cubic metals the metallic bonding has a slight admixture of covalent

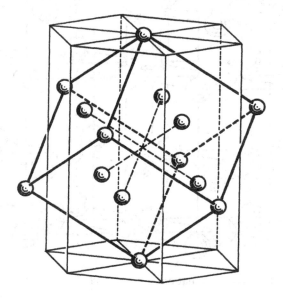

FIG. 5 — Cubical unit cell.

TABLE 6. Crystal Structures of Some Metals

Hexagonal Close-Packed	Face-Centered Cubic	Body-Centered Cubic
Beryllium	Aluminum	Chromium
Cadmium	Calcium	Iron
Cobalt	Copper	Lithium
Magnesium	Gold	Molybdenum
Thallium	Lead	Potassium
Titanium	Platinum	Sodium
Zinc	Silver	Tungsten

bonding. The electron gas probably shows a tendency to collect along lines that join the atoms and make definite angles at each.

Spherical Ions

Thinking about crystals made of more than one kind of atom, William Barlow suspected more than seventy years ago that the atoms in sodium chloride are arranged as shown in Fig. 6. Objections were raised that Barlow's structure could not possibly represent sodium chloride, because his proposal does not portray the atoms as associated in diatomic molecules. But studies of sodium chloride by X-ray diffraction showed that Barlow was right; the structure gives evidence that in solid sodium chloride the atoms are not paired off to form molecules.

FIG. 6 — Barlow's structure of sodium chloride.

In Chapter VI the structure of sodium chloride was used to justify the idea that the sodium and chlorine atoms, in crystals of that substance, are ions bearing net charges. Each sodium ion surrounds itself with six chloride ions, and each chloride ion with six sodium ions; each ion attracts its opposites indiscriminately. Indeed, a molecule of sodium chloride, made of one atom of each species, is a great rarity. Molecules of that sort are found in gaseous sodium chloride, formed at high temperatures, and probably nowhere else.

Barlow's picture of the arrangement of the ions in sodium chloride shows the ions of one element as larger than the ions of the other. Today we know that the chloride ions are larger than the sodium ions. When a chlorine atom becomes an ion by acquiring an extra electron, the added negative charge repels all the electrons that were already in the atom. By trying to stay away from the added electron, all the electrons use up more space and the atom expands. For converse reasons, the sodium atom contracts when it loses an electron and becomes a positive ion.

Barlow placed the spheres representing the larger atomic elements in contact with one another, and probably this is almost true of the chloride ions in sodium chloride. Notice that the negative ions are consequently close-packed, in the face-centered cubic way. It is interesting to compare this fact with the fact that the atoms in many metals are also close-packed in the same way. In those metals, the negatively charged free electrons are holding the larger positively charged metal ions in close-packed array. In sodium chloride, the positively charged ions are holding the still larger negatively charged ions in that array.

But when the positive ions approach the size of the negative ions, the analogy between the close packing of the negative ions in salt and the positive ions in metals breaks down. The larger positive ion is large enough to accommodate more negative ions around it, and it acquires eight rather than six of those ions as nearest neighbors. The ions adopt the arrangement shown in Fig. 7a, named the *cesium chloride structure* after one of its examples.

The sodium chloride and cesium chloride structures are suitable only for crystals containing two ionic elements in equal numbers. When the valencies of the two elements differ, so too must the relative numbers of ions in the crystal. In the mineral fluorite, for example, the formula CaF_2 means that the fluoride neighbors of each calcium ion should be twice as numerous as the calcium neighbors of each fluoride ion. The structure (Fig. 7b) provides eight fluoride neighbors for each calcium ion at the corners of a cube and four calcium neighbors for each fluoride ion at the corners of a regular tetrahedron.

This description of how the ions are located in fluorite suggests a widely useful way of visualizing complicated crystal structures. It is often helpful to think of the structure as built of simple substructures. Each substructure is occupied by a single atomic element or perhaps by a group of atoms that are always associated with one another. Then the whole structure is pictured by allowing the substructures to penetrate one another.

You have already noticed in Fig. 6, for example, that the chloride ions in a sodium chloride crystal are close-packed in the face-centered cubic way. You can think of the sodium chloride structure as built of two face-centered cubic structures—one of chloride ions, the other of sodium ions—that penetrate each other. The cesium chloride structure consists of two interpenetrating simple cubic structures, of the sort shown in Fig. 8. In the arrangement of the ions in fluorite, a simple cubic structure of fluoride ions penetrates a face-centered cubic structure of calcium ions.

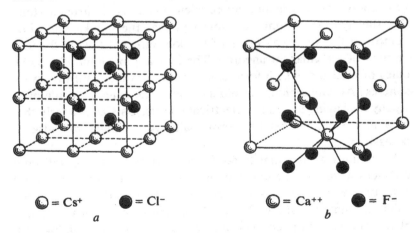

$\bigcirc = Cs^+$ $\bullet = Cl^-$ $\bigcirc = Ca^{++}$ $\bullet = F^-$

a   b

FIG. 7—(a) The cesium chloride structure. (b) The fluorite structure.

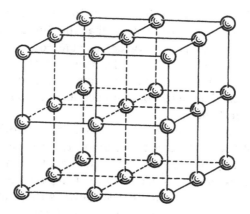

FIG. 8—The simple cubic structure.

The Shapes of Ions

When the ions in a crystal are not formed from single atoms, they will not have spherical shapes. Then the atomic arrangements may be less symmetrical than those formed by spherical ions because they must accomplish a compromise between the influence of those shapes and of the interionic attractions and repulsions.

To be sure, an ion may be so nearly spherical that it behaves like a sphere. The ammonium ion, for example, consists of four hydrogen atoms tightly attached by covalent bonds to a nitrogen atom. But the hydrogen atoms are so small that they make only little bumps on the spherical shape of the ion, directed toward the four corners of a regular tetrahedron. The ions are much like the molecules of methane, which form close-packed crystals as if they were spheres. An ammonium ion differs from a methane molecule principally by having a positive electrical charge. Its size is nearly that of a potassium ion, and it forms salts very similar to those of potassium.

Sometimes an ion that is far from spherical in shape will nevertheless have a high enough symmetry to enable it to form crystal structures closely related to those formed by spherical ions. In the chloroplatinate ion shown in Fig. 9a six chlorine atoms surround a platinum atom at the corners of a regular octahedron. Joining with the ammonium ion, it fits neatly into the structure shown in Fig. 9b.

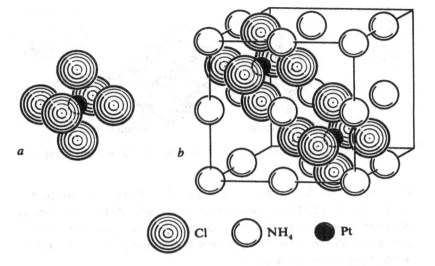

FIG. 9 – (a) Chloroplatinate ions joining ammonium ions to form (b) ammonium chloroplatinate.

You can think of that structure as the fluorite structure, with ammonium ions replacing fluoride ions and chloroplatinate ions replacing calcium ions but with their electrical charges reversed in sign.

When the ions have less symmetry than these, they may form structures that you can picture as distortions of more familiar structures. The nitrate ion, for example, is composed of three oxygen atoms bonded to a nitrogen atom. The centers of the oxygen atoms are at the corners of an equilateral triangle, with the nitrogen atom at the middle. In sodium nitrate the sodium and nitrate ions occupy positions rather similar to those of the ions in sodium chloride (Fig. 10). The planes of the triangular nitrate ions are all parallel, and the ions stretch the structure in those planes. For this reason sodium nitrate crystallizes in rhombohedra, whereas sodium chloride forms cubes.

In the mineral calcite – composing such familiar materials as chalk, limestone, and marble – calcium carbonate adopts the same structure as sodium nitrate. Calcium acts as a stand-in for sodium and carbonate for nitrate. Because their structures are the same,

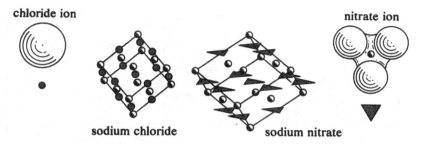

FIG. 10—The structures of sodium chloride and sodium nitrate.

their properties also are very similar. They form a counterpart to diamond and graphite, showing that two solids of different composition may become similar by adopting the same structure, just as two solids of the same composition may become very different by adopting different structures.

Tetrahedral Bonding

In these examples — the rare gases, the metals, the assemblies of ions — the crystal structures can be understood by thinking of their ingredients as attracting or repelling one another by a force that can operate on an indefinite number of them. The van der Waals forces and the ionic forces exhibit no preference for direction and no limit to the number of atoms on which they will operate. But in such metals as iron the body-centered cubic structure gives evidence that directed forces are supplementing the metallic bonds and limiting the number of neighbors nearest to each atom. And within a complex ion, such as the nitrate ion, strongly directional forces with even greater discrimination hold the constituents together, giving the ion its characteristic shape.

Covalent bonds, restricted in number and direction, are responsible for these atomic arrangements. Diamond is a conspicuous example of their operation. Each carbon atom has only four nearest neighbors rigidly placed about it at the four corners of a regular tetrahedron to form the structure pictured back in Chapter IV as Fig. 9a. Surely, if all the atoms were attracting one another indiscriminately, each would acquire more neighbors. The structure is better visualized as a network of bonds than as an assembly of spheres; the network provides atomic cages that leave much open space.

The strong directionality of these covalent bonds is a consequence of the way in which the atoms share their electrons. When there are other atoms about, the orbits that are available to the shared electrons in a carbon atom tend to arrange themselves so that they project along tetrahedral directions. Thus, the shared electrons pull the sharing atoms into tetrahedral positions. For the same reason silicon and germanium—elements that appear beneath carbon in Group IV of the Periodic Table—also form crystals with the same structure.

Notice in Fig. 11 that the diamond structure can be visualized as made of two face-centered cubic substructures that interpenetrate. The two substructures are displaced from each other along the *body diagonal*—the line from corner to corner—of a cubic unit cell. There are many substances that crystallize with a closely related arrangement of their atoms, in which the two substructures are occupied by two different elements.

In boron nitride, for example, each boron atom is tetrahedrally surrounded by nitrogen atoms, and each nitrogen atom by boron atoms (Fig. 12). Its structure exemplifies again the donor-acceptor bond between nitrogen and boron, as described in the preceding chapter, in the molecule constructed from ammonia and boron

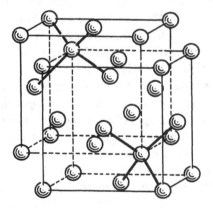

FIG 11—The diamond structure (see Fig. 9a of Chapter IV) formed of two interpenetrating face-centered cubic structures (Fig. 5).

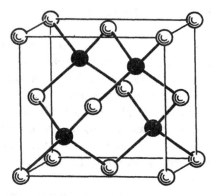

FIG. 12—The zinc blende structure adopted by boron nitride.

trifluoride. In either instance you can think of each nitrogen atom as losing an electron to a boron atom, giving the boron atom a negative charge and leaving the nitrogen atom with a positive charge. Then the nitrogen and boron ions alike have four electrons in their outermost occupied shells, and they share those electrons just as the carbon atoms share their electrons in diamond. Each bond, therefore, has both ionic and covalent ingredients.

In the same way gallium arsenide and indium antimonide form crystals with the same structure. Compounds such as these are often called *three-five compounds* because their ingredients fall in Groups III and V of the Periodic Table. The similarity of their structure to that of silicon and germanium gives many of them similar physical properties; in particular, they are finding use as semiconductors (Chapter XIII).

What happens when you depart still further from the middle of the Periodic Table and make *two-six compounds*? Ambiguities arise, which you can imagine by taking still another step, to *one-seven compounds*. You have already noticed that one-seven compounds adopt the sodium chloride and cesium chloride structures, dominated by the spherical shapes of the ions and the undirected ionic forces between them. As in the three-five compounds, you can picture in the one-seven compounds the loss of one electron from each atom of one species to an atom of the other species. But as Fig. 13 shows, the transferred electron crosses the Periodic

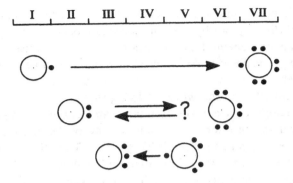

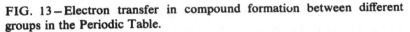

FIG. 13—Electron transfer in compound formation between different groups in the Periodic Table.

Table in opposite directions, so to speak, in the two types of compounds.

Which way will the two electrons take, crossing the Periodic Table to form the two-six compounds? The answer seems to be sometimes one way and sometimes the other. Magnesium oxide forms ionic crystals with the sodium chloride structure (Fig. 6); the electrons go to the right. Zinc sulfide, on the other hand, crystallizes in the mineral sphalerite with the same atomic arrangement as boron nitride (Fig. 13); the electrons go to the left. But whichever way the electrons cross, they will produce doubly charged ions, and those large charges will tend to move the electrons back toward the atoms from which they came. The ions will share their electron pairs unsymmetrically, and the effective charge of an ion will be less than two electronic units.

Rocks

The remarkable ability of carbon atoms to form strong covalent bonds with one another—the ability whose climactic achievement is diamond—accounts for the diversity of organic compounds. Each molecule of those materials is built upon a skeleton of carbon atoms tightly bonded in rings and branching chains. Not even silicon, its closest ally in the Periodic Table, rivals the ability of carbon to proliferate molecular species by bonding to itself in so great a variety of stable skeletons.

But silicon is almost as unusual in another way. It can bond to itself, not directly but by way of an oxygen bridge; and again those bonds can be repeated and extended to form a great variety of structures. The Si—O—Si bond takes much the same part in building skeletons for the molecules of the inorganic world that the C—C bond takes in the world of living things.

The analogy between the inorganic and organic bonds is especially clear in the mineral cristobalite, one of the many crystalline forms taken by silicon dioxide (Fig. 14). There the silicon atoms occupy the carbon sites of the diamond structure, and an oxygen atom falls midway between each pair of silicon atoms. In other words, the oxygen atoms occupy sites at the midpoints of the bonds in the diamond structure. Each C—C bond in the diamond structure is replaced by an Si—O—Si bond in the cristobalite structure.

Notice that you can think of each silicon atom as located at the center of a regular tetrahedron whose corners are occupied by oxygen atoms. In forming cristobalite, these SiO_4 tetrahedra can be pictured as joined by sharing the oxygen atoms at their corners (Fig. 15). Here is a clue to a method for analyzing a great diversity of silicate minerals. Their silicate skeletons are usually formed by SiO_4 tetrahedra, connected so as to share oxygen atoms at some of their corners.

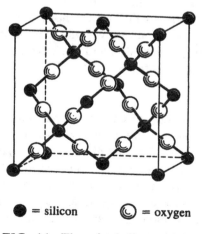

● = silicon   ◐ = oxygen

FIG. 14—The cristobalite structure.

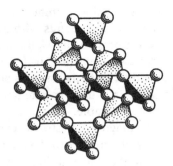

FIG. 15—The cristobalite structure shown as made of SiO$_4$ tetrahedra.

When we look at how these minerals accommodate their other atomic ingredients, it is helpful to compare a silicate skeleton with a corresponding carbon skeleton in an organic compound. Figure 16 shows, for example, a simple chain of linked carbon atoms and the analogous chain of silicate tetrahedra. In forming an organic compound, the carbon skeleton clothes itself by establishing covalent bonds to other atoms with its remaining unpaired electrons. If those other atoms are hydrogen atoms, the product is the modern plastic polythene.

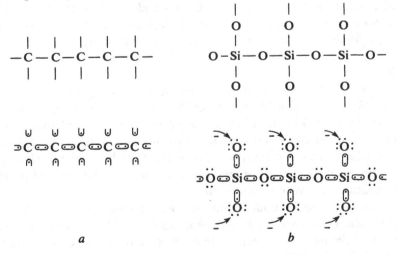

FIG. 16—Comparison of (a) carbon skeletons of organic compounds with (b) silicate skeletons of rocks.

On the other hand, in forming a mineral, the silicate skeleton clothes itself by establishing ionic bonds with metal atoms that lose electrons to the oxygen atoms in the skeleton. Each SiO_3 unit of the skeleton can accept two electrons in this way. The positively charged metal ions then distribute themselves among the negatively charged silicate skeletons in whatever manner their sizes and the ionic forces make most favorable. In the mineral diopside equal numbers of calcium and magnesium atoms furnish the metal ions.

Often the silicate tetrahedra are linked in rings or in sheets instead of in chains. In the mineral benitoite, for example, the tetrahedra form three-membered rings, and each ring is attended by a barium ion and a titanium ion. In beryl a six-membered ring of silicate tetrahedra accommodates three beryllium ions and two aluminum ions.

Clearly the great differences between the soft low-melting crystals of organic compounds and the hard high-melting minerals come principally from the differences in the way their skeletons are clothed. The covalent bonds to a carbon skeleton make the clothed skeleton an electrically neutral molecule, and then the molecules attract one another only by the weak van der Waals forces between them. The ionic bonds to the silicate skeleton make it impossible to distinguish molecular units in the minerals, just as molecules of sodium chloride cannot be distinguished in salt. And the strong ionic forces hold the entire assembly tightly together.

Ice

Silica itself crystallizes in a bewildering variety of structures, but, like the cristobalite structure (Fig. 14), all can be regarded as assemblies of SiO_4 tetrahedra sharing oxygen atoms at their corners. Another of the crystalline forms of silica is tridymite, a mineral whose structure deserves special attention because it is closely related to the structure of ice, one of the most important and most mysterious solids in man's world.

You can easily understand the arrangement of the atoms in tridymite by comparing it with that in cristobalite and imagining it to be built out of interpenetrating substructures. Since the silicon atoms in cristobalite are arranged in the diamond structure, they occupy two interpenetrating face-centered cubic substructures. In tridymite the silicon atoms occupy two hexagonal close-packed substructures. As in diamond and cristobalite, these two substructures are displaced from each other so that each site in either is tetrahedrally surrounded by four sites belonging to the other. And again the neighboring silicon atoms are connected through oxygen bridges.

A good way to construct a picture of the tridymite structure is to build it up in four stages. Two hexagonal close-packed structures

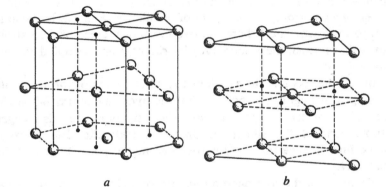

a   *b*

● = silicon
◎ = oxygen

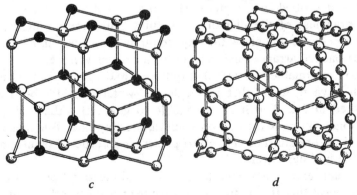

c   *d*

FIG. 17 — The tridymite structure built in four steps.

of silicon atoms (Fig. 17a and b) interpenetrate to form a structure (Fig. 17c) in which each silicon atom is tetrahedrally surrounded by four others. Oxygen bridges then complete the structure (Fig. 17d) in the same way that such bridges complete the structure of cristobalite (Fig. 14).

Thus, in cristobalite and tridymite, respectively, the two simplest close-packed structures — cubic and hexagonal — play the role of substructures. It is remarkable that there is no instance of a single atomic species forming crystals in which the atoms adopt the arrangement of the silicon atoms in tridymite. But zinc sulfide crystallizes not only in the form of sphalerite (Fig. 12) but also in the form of wurtzite, where the two hexagonal close-packed structures that furnish the silicon sites in tridymite are occupied by zinc and by sulfur.

In ice the oxygen atoms adopt the silicon sites of the tridymite structure. It might be supposed that the hydrogen atoms fall at the midpoints of the bridges between the neighbors, just as oxygen atoms connect the silicon neighbors in tridymite. But there is powerful evidence that the hydrogen atoms do not behave in quite that way.

To be sure, the hydrogen atoms are probably located along the lines between neighboring oxygen atoms, but each hydrogen atom is nearer to one oxygen atom than to the other on its line. The hydrogen atoms hold the oxygen atoms together by the hydrogen bonds described in Chapter VII into a cagelike structure having the openness of the diamond structure. When the ice melts, the cages collapse and allow the oxygen atoms to adopt a more nearly close-packed arrangement, occupying a smaller volume.

It is reasonable to guess that only two hydrogen atoms are near each oxygen atom — that molecules of water can still be distinguished in ice. And it is reasonable to guess also that only one hydrogen atom is located on each connecting line. But these restrictions are still not enough to prescribe an orderly arrangement for the hydrogen atoms.

In fact, experiments on the thermal and electrical properties of ice suggest that the hydrogen arrangement is disorderly. A hydrogen atom can jump between the two favored positions along its line, and it can also jump to a neighboring line. If other hydrogen atoms take appropriate jumps at the same time, the conjectured restrictions can still be fulfilled.

This instance of disorder within an orderly structure is unusual but not unique. Discussing the movement of atoms in solids, the next chapter will describe another extreme example.

IX. ATOMIC MOTIONS

The opponents of the atom are generally content to point to its contradictions and reject it as unfruitful for science. A rash form of caution, for without the atom science falls.

HANS VAIHINGER, *"The Philosophy of As If"*

PICTURING the structures of crystals—their orderliness and symmetry, and how the atoms are bonded in them—the last few chapters have portrayed the atoms in fixed positions. But, of course, as the description of heat in earlier chapters emphasized, the atoms are actually in ceaseless motion. The orderly sites pictured in the structural diagrams are only the average positions of the atoms, the positions about which they vibrate. As Dame Kathleen Lonsdale puts it,

A crystal is like a class of children arranged for drill, but standing at ease, so that while the class as a whole has regularity both in time and space, each individual child is a little fidgety!*

Thus, the orderliness of a crystal is at best an average orderliness. At any instant the crystal always has some disorder, since the unorganized vibrations of the atoms displace them from their orderly sites in a random way. In most crystals each atom constantly returns to its own site as it vibrates; the atoms do not interchange sites or stray permanently from their proper places. Heating a crystal increases the vigor of the atomic vibrations until the atoms hit one another hard enough to make spaces in which they can move past one another. Then the crystal melts.

Crystals and X-Rays (Princeton, N.J., Van Nostrand, 1949).

This is the normal behavior of the atoms in a crystal, but there are some interesting exceptions. You have already been introduced to two sorts of motions of atoms that carry them to new crystalline sites. When the lines of dislocations described in Chapter IV are shifted by applying a shearing stress to a crystal, the atoms along those lines glide into new sites. And the hydrogen ions in ice can jump between two oxygen atoms or between two bonds, as the last chapter described.

There is a distinction between these occurrences that suggests a way of classifying atomic motions. Dislocations move under the urging of a directed force applied from outside the crystal. The hydrogen ions in ice jump under the influence of the disorganized thermal vibrations, which occasionally give one atom or another the extra energy required to make it jump. And there are many other instances of both organized and disorganized atomic motions in crystals.

Infrared Absorption

A very important kind of organized atomic motions is induced in a crystal by infrared light. Remember that light is an electromagnetic wave propagated through space at the very high velocity of 186,000 miles per second. The length of the waves in visible light is only a few hundred-thousandths of an inch, varying with the color of the light from about 1.5×10^{-5} inch (385 mμ) for violet light to about 3×10^{-5} inch (760 mμ) for red light. Ultraviolet light has shorter waves and infrared light longer waves, placing them outside the extremes of the light that human eyes can see.

Since the nature of light is electromagnetic, it puts an electrical force on any electrical charge that it encounters. To say that light has, for example, a wavelength of 2×10^{-5} inch is the same as to say that the electrical force of the light reverses its direction every 10^{-5} inch (Fig. 1). A train of waves in which an electrical force constantly reverses direction will urge a charged particle first one way and then the other as it passes the particle. When the force in the waves reverses direction every 10^{-5} inch, and the waves rush past the particle at 186,000 miles per second, the frequency with which the force alternates is very high—nearly 10^{15} times per second.

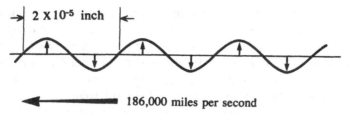

186,000 miles per second

FIG. 1 – A beam of light composed of electromagnetic waves.

Figure 2 shows how a crystal of sodium chloride will respond to that rapidly alternating electrical force. At any instant the force will try to push the positively charged sodium ions in one direction and the negatively charged chloride ions in the other. But the force is alternating so rapidly that before the ions can respond, the force has reversed its direction. Trying to move the ions by that alternating force is like trying to move a child on a swing back and forth by attaching the swing to the clapper of a doorbell.

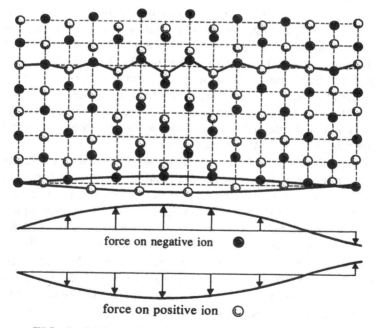

FIG. 2 – Light passing through a sodium chloride crystal.

To swing the child, you must alternate your push at a lower frequency, and to move the ions, you must use light whose frequency is only about 1/100 of the frequency of visible light. At a frequency corresponding with some wavelength of infrared light, you will begin to build up oscillations in which the oppositely charged ions swing in opposite directions. The exact frequency at which this happens will be characteristic of the substance; it will depend on the masses of the ions and the stiffness of the springs in the crystal that restore the ions to their proper sites.

In other words, infrared light of a particular frequency drives one of the harmonic oscillators used in Chapter III to explain the specific heats of solids. By comparing Fig. 9 of that chapter with Fig. 2, above, you can see that the light must drive the oscillator with the highest frequency, in which adjacent atoms move in opposite directions.

This argument suggests that the oscillators of highest frequency, constantly driven by the light, would swing with constantly increasing amplitude until something gave way. But all the oscillators in the crystal interact with one another and constantly exchange energies with one another. When one oscillator gains more than its proper share of energy, it almost always feeds some of that energy to the others. Thus, the energy supplied to one oscillator by the infrared light is soon distributed among all of them. In other words, as the light is absorbed, the temperature of the crystal rises. The organized motion driven by the light is rapidly transformed into the disorganized motion characteristic of heat.

In the experiments demonstrating this absorption, a crystal is exposed to beams of infrared light of known wavelengths, one after another, and the amount of light that the crystal transmits is measured. Plotting the transmission against the wavelength yields a curve like that in Fig. 3. The shortest wavelength for which appreciable amounts of light are absorbed is called the *infrared absorption edge* and corresponds with the frequency of the most rapid structural vibrations in the crystal. Light of the longer wavelengths drives some of the oscillators that have lower frequencies. And as the comparison of Fig. 9 in Chapter III and Fig. 2 above would predict, the frequency of the absorption edge corresponds closely with the *maximum frequency* that appears as the undeter-

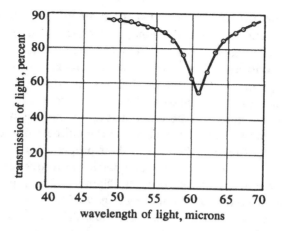

FIG. 3 – Transmission of light through sodium chloride.

mined parameter in Debye's theory of specific heats described in Chapter III.

Diffusion

The atomic motions stimulated in a crystal by infrared light still leave the atoms in the same average positions. Long before the stimulated oscillator can build up enough amplitude to push atoms past one another, the oscillator has delivered its extra energy to the random thermal motions of the structure. Interchanging the positions of two atoms would require very large forces (Fig. 4a).

But there are crystals that give evidence that their atoms occasionally jump to new sites. The atoms move through them slowly, and no doubt they accomplish those travels by making a succession of such occasional jumps. Almost certainly the atoms are able to jump because the crystal has some of the defects described in Chapter IV, especially vacancies.

The presence of a vacancy can enable an atom to jump in the way shown in Fig. 4b. An atom can move into a vacancy instead of interchanging places with another atom. In this action the atom moves the distance between two adjacent sites, and the vacancy moves the same distance in the opposite direction, where it can then enable another atom to move.

Thus, a particular vacancy can move farther than a particular

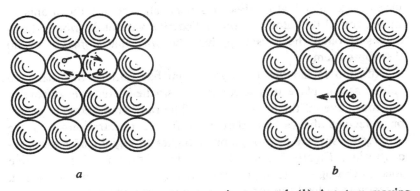

FIG. 4—(*a*) Interchanging two atoms in a crystal. (*b*) An atom moving into a vacancy.

atom in a given length of time. A vacancy moves through the crystal more rapidly than an atom, and the atom must wait between jumps until some vacancy appears next to it. At a higher temperature the crystal has more vacancies. Furthermore, each atom is vibrating more vigorously at a higher temperature and can jump into a vacancy more easily. For both these reasons the rate at which atoms can move through a crystal increases rapidly with increasing temperature.

Ordinarily, of course, there is no way to see from outside that the atoms are moving. The jumping is a random process, in which as many atoms move in one direction as in another. But in recent years it has become possible to study the process by using radioactive isotopes.

Many atomic species have isotopic variants—atoms whose electronic structures are the same and whose nuclei have the same electrical charge but different masses. Since they have the same electronic constitution, the variants behave alike in almost all respects. They have the same size and the same chemical characteristics; they are distinguishable only by the differences between their nuclei. When one of the variants has nuclei that are radioactive, radiation from those nuclei makes their identification easy.

The motion of the atoms through metallic crystals has been studied by plating a thin layer of a radioactive isotope of the metallic species on one surface of such a crystal. Sections of the crystal, examined after time has been allowed for the atoms to move, reveal that atoms of the radioactive variant distribute themselves through the crystal in just the way that *diffusion* would distribute them.

Diffusion is familiar to anyone who has noticed that perfume spilled at one place in a room can be smelled later everywhere in the room even if the air seems still. The random motions of atoms or molecules always tend to even out their concentration — to move a diffusing species from regions of higher to regions of lower concentration. Figure 5 suggests how the diffusion of an isotopic species through a metal probably proceeds. Any atom that faces a vacancy is as likely to jump as any other. But when there are more atoms of one isotopic species on the left than on the right, more will jump from left to right than from right to left because there are more on the left to make the jump.

Color Centers

A similar process can be made to occur in some ionic crystals. When a crystal of sodium chloride, for example, is heated in a closed vessel that contains a vapor of metallic sodium, a red-brown color slowly diffuses into the crystal. There is much evidence that the color is due to electrons that are trapped in vacancies.

Here it is not necessary to suppose that the vacancies were present before the process began. They can be created by the process itself. When a sodium atom in the surrounding vapor comes to the surface of the sodium chloride crystal (Fig. 6a), a chloride ion can jump out to join it. By losing one electron, the sodium atom becomes a sodium ion just like the others in the crystal. The lost electron will go into the vacancy (Fig. 6b) left at the surface by the chloride ion because the negative charge of the electron will be attracted there by the surrounding positive ions.

Thus, the electron enters the crystal by changing places with the chloride ion. Then, by continuing to change places with other chloride ions, the electron can diffuse deeper into the crystal (Fig. 7). The trapped electron can interchange sites with a chloride ion, as another chloride ion could not, because the electron is very much smaller.

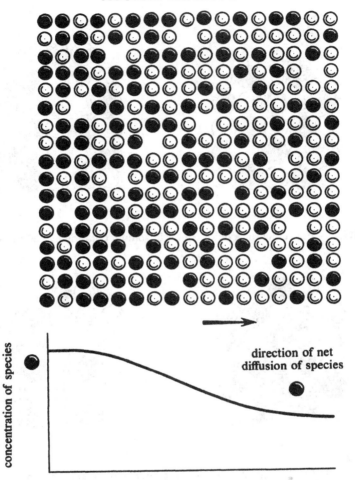

FIG. 5 — A diffusing species of atoms moving from regions of higher to regions of lower concentration of the species.

Electrical Conduction in Ionic Crystals

Ionic crystals, like metals, probably contain many vacancies left in their structure when they grew. The presence of such vacancies would explain why an ionic crystal is never a perfect electrical insulator.

The mechanism by which vacancies permit an ionic crystal to conduct a tiny trickle of electric current is suggested in Fig. 8. A

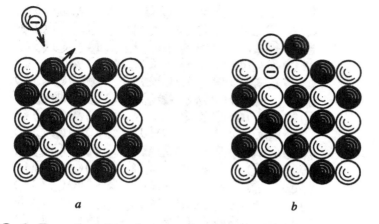

FIG. 6—Formation of a color center in sodium chloride by sodium vapor.

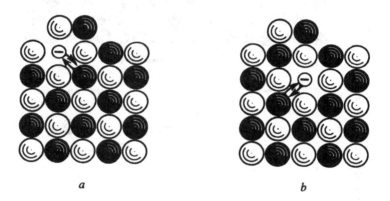

FIG. 7—Diffusion of a color center into sodium chloride.

crystal of sodium chloride probably contains vacancies of two types, *chloride vacancies* and *sodium vacancies*. Both types of vacancies diffuse through the crystal, and the ions move from site to site in the way already described. In the absence of any directed forces, both sorts of ions move in all directions—as many in one direction as in another. But an electrical force applied from outside gives the ions a slight preference to jump in the direction favored by that force.

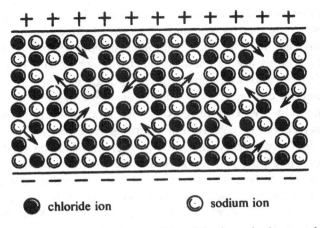

● chloride ion   ◎ sodium ion

FIG. 8—Ionic conduction of electricity in an ionic crystal.

The preference is only slight; the ions still jump mostly at random. The applied force nudges them only a little, so that each ion makes just a few more jumps in one direction than in the other. But the thermal motions of the ions are still primarily responsible for the fact that they jump at all. The electrical conduction in an ionic crystal, therefore, varies greatly with the temperature in the same way that diffusion varies.

It is interesting to translate some of the last chapter's discussion of ice into the terms used in this discussion of diffusion. The alternative positions suitable for occupancy by hydrogen atoms, located in pairs along lines between oxygen atoms, are all hydrogen sites, and half of those sites are vacant. Thus, there are as many vacancies as hydrogen atoms. A hydrogen atom in ice always finds vacancies facing it and need not wait for a vacancy to appear in its neighborhood before it can jump.

Protonic Motion in Solids

The structure and behavior of ice make conspicuous the unusual properties of hydrogen. Hydrogen is unusual in two respects: It is the lightest kind of atom; and when it loses its one electron, the positive ion that remains has no electrons left and is therefore a bare proton, very much smaller than any other common ion. In fact, a proton hardly differs from an electron in size. The important differences are that the proton's charge is positive instead of negative and that it is about 2,000 times heavier than the electron.

Since protons are as small as electrons, why can't they stream through solids as electrons stream through metals and conduct large electric currents? You will find it an illuminating exercise to examine how the two properties just mentioned prevent using protons as electrons are used in electrical technology.

The first obstacle is the much greater mass of the proton. An electrical field puts just as large a force on the proton as on the electron—in the opposite direction, of course, because the charges on the two particles have the same magnitude and opposite signs. But that force will accelerate the proton only 1/2,000 as much. If you managed to put protons in situations comparable to those of the electrons in metals, the electrical resistance would be 2,000 times as large, for that reason alone.

Even greater difficulties, however, come from the opposite charge of the proton. As a consequence of that opposite charge, putting protons in situations comparable to those of electrons can only mean reversing the signs of all the other charged particles. But you cannot make materials out of atoms in which positively charged particles circulate about negatively charged nuclei, because our world contains no such atoms. The resulting difficulties can be understood by recalling the discussion of polarization in Chapter VII.

As any electron moves through the collection of positive ions left behind by the free electrons in a metal, it must pass close to one positive ion after another. While it is close to an ion, it attracts the positively charged nucleus and repels the cloud of bound electrons still circulating about that nucleus. In other words, it polarizes that ion, much as a negative ion and a positive ion polarize each other in an ionic crystal. By repelling the clouds of electrons around the positive ions, the moving electron digs a hole for itself as it goes, easing its passage through the solid.

The proton-conducting solid, on the other hand, must be made of protons and negative ions in order to be electrostatically neutral. A proton moving past a negative ion attracts the cloud of electrons around the nucleus of that ion. It moves much closer to the edge of that cloud than a free electron does to the nucleus of a metal, and it can attract the edge of that cloud quite strongly. Hence, the proton tends to get stuck to the cloud — stuck tightly enough to form a fairly strong bond with the negative ion, as the following example shows.

Ammonium Dihydrogen Phosphate

Ammonium dihydrogen phosphate is a salt whose crystals conduct electricity by the motion of protons. When a direct-current voltage is applied to electrodes placed on opposite surfaces of a single crystal, hydrogen is liberated at the negatively charged electrode. The conductivity of the crystal is more than a hundred times greater than that of most ionic crystals, but many million times less than that of a metal.

The chemical formula of the salt is $NH_4H_2PO_4$; it is composed of positive ammonium ions (NH_4), negative phosphate ions (PO_4), and protons (H_2). Each ammonium ion consists of four hydrogen atoms covalently bonded to a nitrogen atom. Each phosphate ion consists of four oxygen atoms covalently bonded to a phosphorus atom. A phosphate ion carries the extra negative charge of three electrons, one of which it obtains from the ammonium part of the structure and two from the two other hydrogen atoms that furnish in this way two conducting protons.

By looking at the crystal structure of ammonium dihydrogen phosphate (Fig. 9), you can trace with some confidence the tortuous path of a proton from one sticking place to the next. In Fig. 9 the structure is built up in four steps. First visualize (Fig. 9a) evenly spaced strings of ions that are alternately positive and negative and evenly spaced along each string. Then visualize the structure of the gem stone zircon (Fig. 9b), in which zirconium atoms furnish the positive ions and silicate (SiO_4) tetrahedra furnish the negative ions. The structure of ammonium dihydrogen phosphate (Fig. 9c) closely resembles that of zircon, with ammonium (NH_4) ions on the zirconium sites and phosphate (PO_4) ions on the silicate sites but turned a little about the direction of the ionic strings.

The protons provide bonds (Fig. 10) that connect oxygen atoms in neighboring phosphate ions. As in ice, each proton is on a line

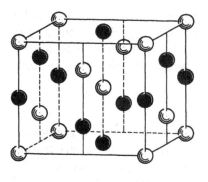
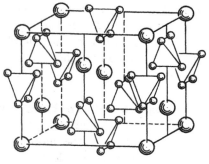

a   *b*

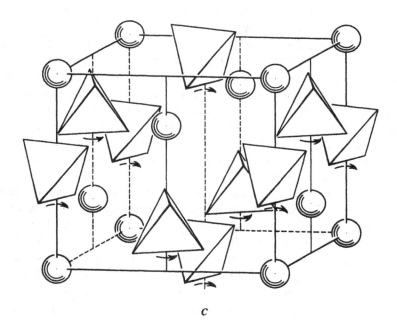

c

FIG. 9—The crystal structure of ammonium dihydrogen phosphate.

connecting two oxygen atoms; again it can take either of two positions on that line and can jump from one to the other. Whereas in ice there are four such lines connected to each oxygen atom, here there are four lines connected to each phosphate ion, one to each of its oxygen atoms. In order to jump from line to line, a

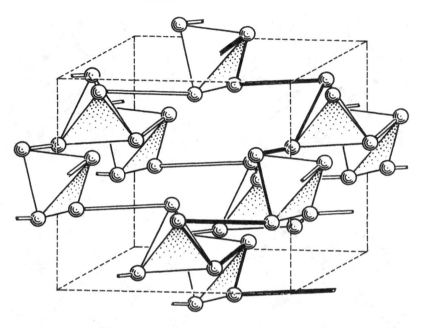

FIG. 10 — A possible path of a proton moving through a crystal of ammonium dihydrogen phosphate.

proton must jump farther here than in ice, but apparently it can still make those jumps. Thus, here again electrical conduction can occur by the motion of protons that take jumps of two sorts.

This picture of electrical conduction in ammonium dihydrogen phosphate makes clear how certain properly chosen ions, entering the crystal as substitutional impurities, increase its conductivity. A sulfate ion, for example, has nearly the same size and shape as a phosphate ion. It contains four oxygen atoms arranged tetrahedrally around an atom of sulfur instead of phosphorus. But the sulfate ion tends to acquire only two extra electrons, one less than the phosphate ion.

Hence, a sulfate ion substituted for a phosphate ion in the crystal carries one fewer hydrogen atom with it, and one of the sites normally occupied by a proton is left vacant. A nearby proton associated with a phosphate ion can move into this additional vacant

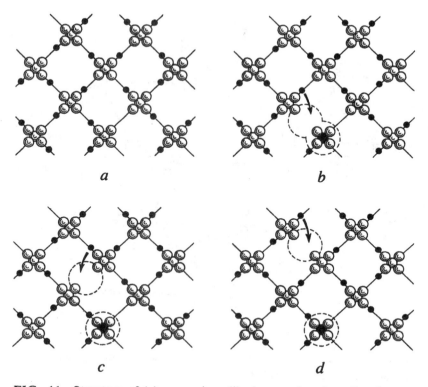

FIG. 11 — Structure of (*a*) ammonium dihydrogen phosphate (*b,c,d*) containing a sulfate impurity.

site, but then it leaves vacant the site that it has deserted. Thus, the improved conduction occurs as in Fig. 11, where the structure of ammonium dihydrogen phosphate is schematized in two dimensions (Fig. 11*a*), omitting the ammonium ions. When a sulfate ion is incorporated, it provides a proton vacancy (Fig. 11*b*), and proton can replace proton successively (Fig. 11*c* and *d*).

Another impurity producing the same result is barium. A barium ion has nearly the same size as an ammonium ion. Since barium has valency 2, each atom of barium will contribute two electrons to a phosphate ion, in contrast to ammonium's contribution of one, and the acquisitiveness of the phosphate ion can then be satisfied by one hydrogen atom instead of two. Again an additional vacancy is left; again, urged by an electrical field, a nearby proton can move

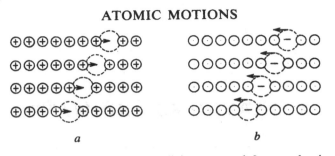

FIG. 12—Two ways of visualizing proton-defect conduction.

into the additional vacancy and be replaced in turn by the motion of another proton.

More simply, however, you can think of the added conduction contributed by the easier motion of these protons as contributed instead by the motion of the additional vacancy in the opposite direction. Figure 12 contrasts the two ways of looking at such conduction. When enough protons have moved to shift the vacancy away from the parent impurity ion, the structure of the crystal around the vacancy is the same as the structure everywhere else, except that one proton is missing. The removal of its positive charge has left a net negative charge in the neighborhood of the additional vacancy. The compensating positive charge that keeps the whole crystal neutral is frozen in place far behind on the parent impurity ion.

Hence, the motion of the additional vacancy corresponds to the motion of a negative charge, which is equal and opposite to the positive charge on a proton in a direction opposite to the motion of the protons. If you think of the vacancy as a negatively charged particle moving in a matrix of impurity ions that are fixed in place and positively charged, your picture will be reminiscent of the mobile negative electrons moving among the positive ions in a metal.

Here, however, the metal is extremely dilute; there are relatively few free charges, and those charges are much less mobile than the electrons in a true metal. In a later chapter you will see that ammonium dihydrogen phosphate bears a closer analogy to the electronic semiconductors widely used today in rectifiers, transistors, and solar batteries.

X. PARTICLES AND WAVES

Obviousness is always the enemy to correctness.
BERTRAND RUSSELL,
Mathematics and the Metaphysicians

FOR DISCUSSING most of the physics of solids so far, models made of indivisible atoms have served quite well. Heat capacities could be explained by thinking of those atoms as simple mass-points, connected by weightless springs. In order to explain the structures of crystals and the motions of their atoms, we needed to suppose that the atoms have different sizes and different bonding properties, but mostly it was still possible to think of the atoms as the ultimate units for constructing the solid.

The discussion of their bonding, however, invoked the fact that the atoms are themselves complex, consisting of nuclei and electrons. In the picture of the covalent bond, the atoms are putting electrons out toward one another. The picture of metals gives the electrons and nuclei especially great independence. And unfortunately both pictures bring into view some difficulties that cannot be ignored.

A few of those difficulties are so great that they seem intolerable. The model of metallic bonding is inconsistent with the model of heat capacity. The model of covalent bonding in a crystal is inconsistent with that of metallic conduction. Models that are inconsistent with one another cannot be tolerated simultaneously.

In order to get a clear view of the inconsistencies that are at issue, let us examine further those two simple examples. The first example raises a question about how the free electrons in a metal behave when it is heated. The second example questions the arithmetic used for deciding whether a solid will contain free electrons at all.

The First Inconsistency

For the first example, compare the earlier discussion of heat capacity with that of the metallic bond. In Chapter II a model in which each atom is completely free gave a good description of the heat capacity of a gas and explained the measured value of 3 calories per gram-atomic weight. Again in that chapter, a model in which each atom is a harmonic oscillator gave a good description of the heat capacity of a metal, and explained the measured value of 6 calories per gram-atomic weight. But in the model of metallic bonding in Chapter VII, an electron was detached from each atom, and the metal became a solid that still contained just as many atoms and also contained an electron gas.

When these two models are taken together, they predict that the heat capacity of a metal should have a Dulong-Petit value of 9 calories—6 for the atoms plus 3 for the electron gas—instead of the measured 6. It is not acceptable to require free electrons for metallic bonding and electrical conduction and then to ignore them in picturing the absorption of heat without finding some explicit justification.

It is tempting to seek that justification by appealing somehow to the unusually small mass of the electron—only 1/2,000 of the mass of the lightest atom. To be sure, the heat capacity of an atomic gas does not depend on the masses of the atoms; we cannot argue on the basis of mass alone that an electron gas should behave differently from an atomic gas. But we might try to argue that the electrons are really bound in place in a metal by forces that, although they are quite weak, are still strong enough to make harmonic oscillators (Chapter II) out of the electrons. Then the small mass of an electron would tend to give the oscillators a high frequency —high enough perhaps to give a very large value to the quantum of energy $h\nu$ that an oscillator will accept. The resulting quantum effects could reduce the heat capacity as Chapter III described, even at room temperature.

But the electrons move so freely when they conduct electricity through a metal that any force trying to restore them to fixed positions must be very small. A restoring force as small as that would tend to give the oscillators a very low frequency and so to counteract the effect of the small mass.

In order to preserve the small mass and large restoring force required for an oscillator of very high frequency, we might imagine that the electrons themselves provide the restoring force. Since every electron repels every other, the free electrons tend to stay out of one another's way; two electrons approaching each other will exert a strong force pushing each other back. These forces would not interfere with electrical conduction if all the electrons drift in the same direction together to provide an electric current.

But such an argument would overlook the fact that in a metal the electron gas embeds an array of positive ions. Each electron is not only repelled by the other electrons but also attracted by the ions. Hence, the net electrostatic force on an electron is small except when it gets very close to an ion or to another electron, and that net force is the restoring force. The failure of the electron gas to contribute to the heat capacity of a metal cannot be resolved by such arguments as these.

Other Inconsistencies

The second example of an inconsistency appears when we apply to sodium chloride the arithmetic of electron-deficient bonding, described in Chapter VII. In a crystal of sodium chloride each atom has 6 nearest neighbors of the other species. The arithmetic gives 3 bonds per atom. A sodium atom has 1 electron in its outermost occupied shell, and a chlorine atom has 7. The arithmetic gives 4 electrons per atom. Hence, there are $4/3$ electrons per bond. Common salt, with an electron deficiency of $2/3$, should be a metal — a shiny conductor of electricity, not the white insulating grains that season our food.

It will not avail to say that sodium chloride is an ionic crystal, and the arithmetic was designed to deal only with covalent and metallic crystals. The discussion of mixed bonding in Chapter VII emphasized that there is no clear borderline between ionic bonding on the one hand and covalent and metallic bonding on the other. The interatomic bonding in most crystals is a mixture containing these ingredients in different proportions.

A third inconsistency, even more deep-seated, hides in the discussion in earlier chapters. There the electrons in an atom have been pictured as negatively charged particles circulating in orbits about a positively charged nucleus. An electron must traverse its orbit very rapidly in order to avoid falling into the nucleus, much as our man-made satellites must traverse their orbits rapidly in order to avoid falling back to the earth. But, as the discussion of X-ray diffraction emphasized in Chapter IV, electrical charges radiate electromagnetic waves when they move. Since those waves carry energy away from their source, an electron should lose its kinetic energy, slow down, and fall into the nucleus. In fact, however, the atoms about us and within us do not behave in that way; they seem to endure unchanged forever.

They are unchanged, at any rate, unless they are made to take part in chemical reactions, are bombarded with other energetic particles, or are exposed to light of carefully selected wavelengths. And their changes in these circumstances seem just as puzzling as their unchanged endurance in ordinary circumstances. When sodium metal burns in chlorine gas—in other words, when electrons fly from sodium atoms to chlorine atoms—a great deal of energy is evolved; but after that chemical reaction is over, the ions newly formed in the sodium chloride are just as enduring as their parent atoms were.

During the second quarter of this century, these and kindred puzzles were resolved by a radically new theory of electronic behavior. Three ingredients of this theory can be distinguished. The first is the use of the *wave mechanics* invented by Erwin Schrödinger. The second is the idea of the electron's *spin,* introduced by George Uhlenbeck and Samuel Goudsmit. For the third ingredient, aspects of the other two were combined in the *exclusion principle* by Wolfgang Pauli. In order to understand the way by which contemporary physics describes the behavior of electrons in atoms, molecules, and solids, you must understand these three properties of the electron. The rest of this chapter will introduce you to some of the ideas of wave mechanics; spin and the exclusion principle will appear in the following chapter.

Wave-Particle Duality

Wave mechanics is the farthest-reaching of these three novel ideas. It is really an extension of the more familiar principles of mechanics developed by Sir Isaac Newton, which explained so successfully the motion of the planets about the sun — an extension because it includes Newton's mechanics. From wave mechanics the familiar results can still be deduced for objects of familiar size, observed for familiar periods of time. In talking about small objects, however, its assertions begin to depart from those which the older mechanics would make.

The discussion of the heat vibrations of the atoms in a solid has made a similar departure. As Chapter III pointed out, the quantum theory restricts the energies permitted both to a vibrating atom and to a swinging pendulum. But the restrictions on the atom have conspicuous consequences, and the restrictions on the pendulum do not, because the scales of size appropriate to those two objects are very different.

A disconcerting feature of wave mechanics is that it dooms to failure any effort to picture the electrons in an atom with the precision that can be given to a picture of planets circulating about the sun. A particle as small as an electron, inhabiting as small a space as an atom, cannot be followed so precisely as a planet. There is an ambiguity of location of the electron.

That ambiguity can be related to another — an ambiguity in the nature of the electron. Experiments showed that an electron behaves not only as a particle but also as a wave. Indeed, in similar experiments other small particles — protons, neutrons, and even entire atoms — behaved like waves. For example, all these particles can be diffracted by crystals, much as X rays are diffracted in the experiments described in Chapter IV. Diffraction is characteristic of waves and is hard to visualize as a performance by particles.

Even earlier the same ambiguity appeared in some of the behavior of light. Regarded as a wave ever since the eighteenth century, light was found to exhibit particlelike behavior in some experiments. Taken together, the experiments on particles and on waves suggested a deep-lying symmetry in Nature; particles and waves share some properties. The name *wave-particle duality* was coined to describe this symmetry.

In the case of light, most people confidently speak of the wave as an electromagnetic wave, like a radio wave but with a much shorter wavelength. The associated light particle, called a *photon*, is not so confidently visualized, perhaps only because it is a newer idea. Conversely, most people have little difficulty visualizing the electron as a bit of stuff, while the nature of the associated wave seems rather obscure.

Instead of beating a baffled retreat in the face of such ambiguities, it is better to step back for a moment to examine broadly the nature of physical knowledge. Recall first that we have no way to decide what anything is, but only to describe how it behaves. We have become accustomed to describing the behavior of matter in terms of the particles of which it is made and to think of a particle as something that behaves in a way described by Newton's laws of motion. In fact, that description is the only definition that we really have for a particle.

Finding that a thing we have been accustomed to calling a particle behaves, under some newly investigated conditions, in a way not predicted by Newton's laws, we have two alternatives. One alternative is to say that the thing is ambiguous; under some circumstances it is truly a particle and under others a true wave. Soon after the early experiments in which particles were diffracted by crystals, many physicists seemed to adopt that alternative. It was jokingly said sometimes that a particle behaves like a particle on Mondays, Wednesdays, and Fridays and like a wave on Tuesdays, Thursdays, and Saturdays.

Today physicists prefer the other alternative. They find that they can describe all the behavior of particles by treating them as if they were waves. The newer mode of description, by wave mechanics, describes the behavior that Newton's laws describe correctly and also the wavelike behavior that Newton's laws do not describe. A particle can now be defined as a thing that obeys wave mechanics in such a way that it also obeys Newton's laws under all the appropriate circumstances.

Picturing Particles

Fortunately, wave mechanics still permits you to picture particles as bits of stuff, if you paint waves into your picture along with them. For example, you must picture an electron, when it is moving freely in space, as belonging to a wave-packet traveling in the way suggested in Figs. 1 and 2. Somewhere in this wave is the particle. Where there is no wave, there is never a particle; where there is never a particle, there is no wave. The height of the wave at any point measures the probability that you would find the particle at that point. In fact, the square of the height of the wave gives that probability exactly.*

Here is an answer to the first question that naturally arises: What are the waves that accompany the particle? Recall again that we can answer what they are only in terms of what they do; and what they do is to determine probabilities. They are *probability waves.*

This is the first important idea of wave mechanics. By calculating the waves belonging to particles under different circumstances, you can learn the probability for finding the particle in different regions of space. Studying Figs. 3, 4, and 5 will set you on the road to an understanding of what must be pictured.

Figure 3 displays a small part of the wave of Fig. 1, stretched out (Fig. 3a) along the line of travel. The square of the wave (Fig. 3b) is everywhere positive because the square of a negative quantity is a positive quantity. That square measures the relative probability of finding the particle at different points in the wave. Since the wave is moving, those probabilities are moving also, reflecting the fact that the particle is moving. The probabilities can also be diagrammed by a frequency distribution, or bar chart (Fig. 3c), or by taking samples. In the sampling diagram (Fig. 3d) the size of each

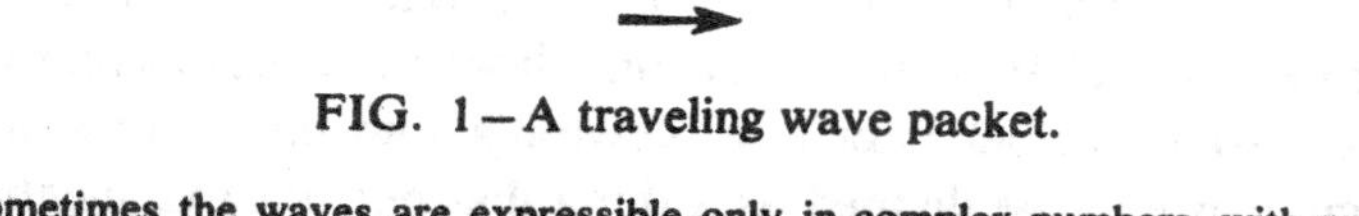

FIG. 1 – A traveling wave packet.

*Sometimes the waves are expressible only in complex numbers, with real and imaginary parts, and in such cases the probability is found by multiplying those numbers by their complex conjugates. When the numbers have no imaginary parts, that formula reduces to taking the squares of the real numbers.

a

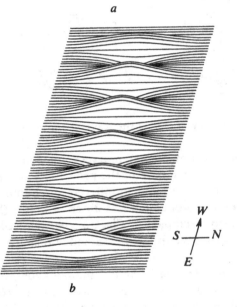

b

FIG. 2—Two imagined views of the wave packet belonging to a particle moving freely from west to east.

sphere is proportional to the relative probability of finding the particle at the point about which the sphere centers. Such a diagram becomes the only method available for showing in a single picture a probability distribution in three dimensions, such as we will soon examine.

Figure 4 shows how the two-dimensional wave of Fig. 2 can be diagrammed in ways analogous to the diagrams in Fig. 3 for the one-dimensional wave. Again you see a small section of the wave (Fig. 4*a*), its square (Fig. 4*b*), and samples (Fig. 4*c*) of the relative probabilities.

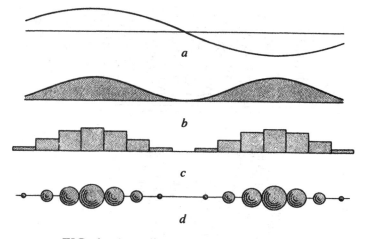

a

b

c

d

FIG. 3—A small part of the wave of Fig. 1.

In Fig. 5 these methods of diagramming are applied to a three-dimensional wave. Here the square of the probability wave can be shown only by taking cross sections in different directions (Fig. 5*a* and *c*) or by sampling the probabilities at different points (Fig. 5*b*). For the probability wave sampled in Fig. 5*b*, the shape of a cross section along the wave is shown in Fig. 5*a* and across the wave in Fig. 5*c*.

Wave mechanics provides mathematical methods for calculating the shapes and motions of the probability waves, just as Newton's mechanics provided methods for calculating the positions and motions of bits of stuff. The new methods, like the older methods, are presumed to be applicable to any sort of object or particle, and they have been successfully applied to many things other than electrons. The place of the electron in the wave-mechanical scheme is merely that of a thing with a certain mass and a certain electrical charge.

In consequence of its negative electrical charge, an electron in an atom is attracted to the atom's positively charged nucleus, and the probability at any instant that the electron is near the nucleus is higher than that it is far away. The square of its wave might look like Fig. 6, peaked up near the nucleus. Here you see an instance of how the first important idea of wave mechanics operates. The

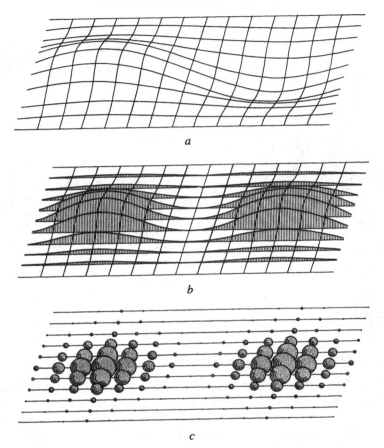

FIG. 4—A small part of the wave of Fig. 2.

picture of an electron in an atom is no longer that of a bit of stuff circulating in a well-defined orbit about the nucleus. Instead, the picture portrays the electron as a probability wave that is clustered about the nucleus and is standing still. Like the preceding illustrations, Fig. 6 shows a cross section of the wave (Fig. 6a) and of its square (Fig. 6b) and a three-dimensional display of probability samples (Fig. 6c).

Imagine, if you like, that a bit of stuff is whirling about in that stationary wave and that the wave merely tells you the relative amounts of time that the bit of stuff spends in one place and an-

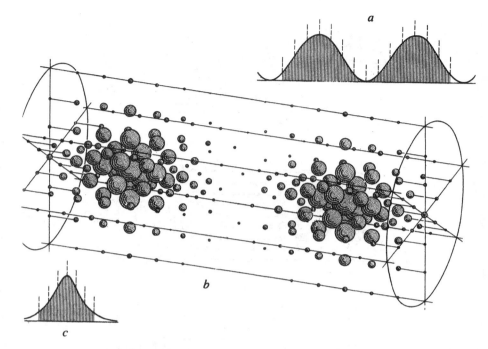

FIG. 5 — Probability samples of a wave in three dimensions.

other. But notice in any case that wave mechanics provides no suggestion of an experiment by which to determine the path that the bit of stuff takes. The probability wave, together with formulas for using it, completes the knowledge that wave mechanics provides. And perhaps the idea that the probability wave is standing still will help you to picture why the electrically charged particle is not radiating its energy into an electromagnetic wave.

The de Broglie Wavelength

The second important idea of wave mechanics connects the velocity of the particle with the length of its wave. The connection is made by a relationship suggested by Louis de Broglie in 1923, which formed a cornerstone for the structure of wave mechanics.

De Broglie's equation makes the *momentum* of a particle inversely proportional to its wavelength. The momentum of a particle is the product of its mass times its velocity; accordingly, the faster

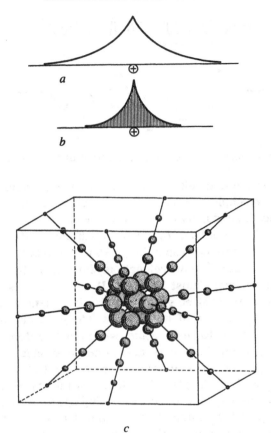

a

b

c

FIG. 6 — A standing wave for an electron in an atom.

a particle is moving, the shorter its wavelength will be. The waves of a rapidly moving electron and a slowly moving electron might offer the contrast shown in Fig. 7. The equation is $mv = h/\lambda$, where mv is the momentum, λ is the wavelength, and h is Planck's constant, to which you were introduced in the discussion of the heat capacities of solids.

From these two ideas — first, the connection between the probability of finding a particle in a place and the square of the magnitude of its wave at that place and, second, the connection between the velocity of a particle and the length of its wave — you can

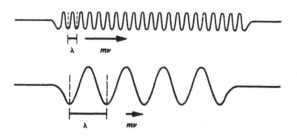

FIG. 7—Wavelength (λ) and momentum (mv) of fast and slow particles.

easily derive some qualitative pictures of the behavior of atoms. Notice, for example, a way of thinking about electrons in atoms which comes immediately out of these ideas.

If a particle is confined to a very small space, then the length of its wave must be correspondingly short, in order to fit into the space (Fig. 8). But if its wavelength is short, then, by de Broglie's equation, its velocity must be great. If its velocity is great, then its kinetic energy must be great, because its kinetic energy is proportional to the square of its velocity. Therefore, it takes energy to confine a particle to small space; work must be done on it.

In thinking about how work gets done on an electron, giving it kinetic energy when it is in an atom, recall the behavior of a pendulum discussed in Chapter II. When you pull aside a pendulum bob in order to start the pendulum swinging, you do work on it. You work against the force of gravity in order to raise the bob, and the work that you do gives the pendulum potential energy. When

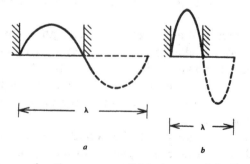

FIG. 8—Decreasing the space available to a particle (from a to b).

you let go of the pendulum, the gravitational force pulls the pendulum down, and it gains speed. Its potential energy is transformed into kinetic energy.

Imagine now a similar performance upon an atom. When an electron is pulled out of it, work is done against the electrostatic attractive force exerted by the positively charged nucleus on the negative charge of the electron. When the electron is freed to return, it moves back toward the nucleus, gaining speed as it goes. The closer it gets to the nucleus, the greater its speed. In this case again, potential energy is transformed into kinetic energy (Fig. 9).

In other words, the kinetic energy needed to give an electron a wavelength short enough to fit into an atom is taken from the potential energy that the electron had when it was farther away. So long as an atom is not in the process of changing in some way, its electrons are said to be in *stationary states;* they are neither losing nor gaining total energy. And the kinetic parts of their total energies determine the wavelengths of the electrons.

Stationary States

Now look for a moment at some consequences of the idea of a stationary state—such a state as an electron has when it is in an unchanging atom. If an electron is in a stationary state—neither gaining nor losing energy—the form of its wave is not changing. A little thought about this fact is sufficient to bring out one of the most curious and important results of wave mechanics.

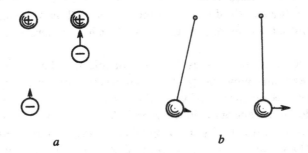

FIG. 9—Comparison of the velocities of (a) an electron in an atom and (b) a pendulum.

The result is that when any particle is not free to roam but is bound by attractive forces or confined to some limited space, the particle is limited in the choice of total energies that it can have. An example of this result already appeared in the discussion of heat capacity in Chapter III. The harmonic oscillator—a particle bound to its rest position by a spring—can have only energies chosen from a definite set in which two neighboring energies differ by $h\nu$. Similarly, an electron bound by the attractive force of a nucleus in an atom can have an energy chosen only from a definite set.

An easy way to see how this comes about is to look at an especially simple though rather artificial example of a confined particle. Imagine that a particle is strictly required to move in a circle—that it is confined to the inside of a tube of very narrow bore which is bent into a circle and joined at the ends. All parts of the circle look alike to the particle, and all that is known about the particle is that it is somewhere on the circle; so long as the particle stays on the circle, it is free to move.

Then its probability wave is a simple wave, like those pictured for perfectly free particles in Figs. 1 and 7, and retains the same form all the way around the circle. But in order to retain that form after the wave has gone once around the circle, it must join smoothly onto itself (Fig. 10). Clearly it can do this only if its length fits evenly into the circle. In other words, the only waves permitted are those whose lengths are integral subdivisions of the length of the circumference of the circle. If R denotes the radius of the circle, the length of its circumference is $2\pi R$, and the lengths of the permitted waves are given by $\lambda = 2\pi R/n$, where n is any integer.

The fact that the waves for the particle can have only these lengths explains why the particle can have only certain energies. Since there is no force acting on the particle so long as it stays on the circle, all its energy is in the kinetic form. The permitted values of its wavelength lead, by de Broglie's relation, to permitted values of its velocity and so to the permitted values of its kinetic energy (Fig. 11).

The probability waves for an electron in an atom obey similar principles, but they are somewhat more complicated. By discussing

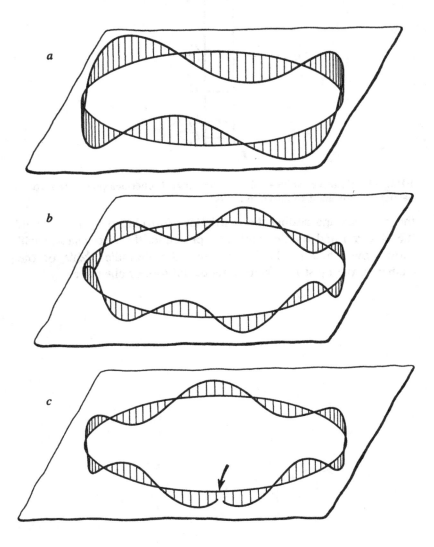

FIG. 10 — (*a* and *b*) Waves permitted and (*c*) a wave not permitted to a particle free to move on a circle.

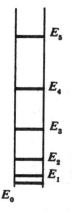

FIG. 11—Relative values of the six lowest energies permitted for a particle constrained to move in a circle.

those waves and adding the restrictions introduced by the spin of the electron and by the exclusion principle, the next chapter will show how wave mechanics explains the Periodic Table of the elements, the most important intellectual tool of chemistry.

XI. ELECTRONS IN ATOMS

The introduction of integers arises in the same natural way as,
for example, in a vibrating string, for which the
number of nodes is integral.
ERWIN SCHRÖDINGER, *Collected Papers on Wave Mechanics*

AN ATOM possesses no tube of narrow bore bent into a circle to confine its electrons. It holds them by much less stringent means—by the attraction of its positively charged nucleus for their negative charges. The electrons are free to escape and to roam through space. Accordingly, the probability waves for electrons in atoms always show some probability—to be sure, a very tiny probability—that any one of them will be found at a great distance from the atom.

But most of the probability for an electron that is in a stationary state in an atom will be found close to the nucleus of the atom. In fact, many of those states turn out to look rather like orbital tracks, smeared out by some uncertainty. You can think of the electron as moving in an orbit during one passage around the nucleus and traversing that orbit so rapidly that the electron is spread out into a streak—much as a rapidly moving source of light can look like a streak—and then moving in a slightly different orbit during its next passage around so that the successive streaks do not quite coincide.

If all this happened before your eyes so fast that the succession of streaks merged into a smear, the density of the smear at different places would be proportional to the amount of time that the electron had spent there. In other words, variations in the density of that smear from point to point would form a picture of the variations in probability that you would find the electron at those various points if there were any way to hunt it down. Since there isn't, you can just as well think of the electron as consisting of the smear that it makes – or, better yet, as consisting of the wave whose square measures the density of that smear.

Energies and Wavelengths

Again, therefore, to describe the electron, you must describe its wave. Just as the permitted waves for a particle confined to the closed tube of the last chapter described the particle's behavior in that tube, the waves permitted to an electron in an atom describe atomic behavior. Again it is helpful to start by thinking of an electron as a bit of stuff.

In the closed tube no forces operated on the bit of stuff so long as it stayed in the tube. Its energy was all in kinetic form. Its permitted wavelengths were determined simply by the length of the tube; then its permitted velocities could be calculated from those wavelengths and its permitted energies from those velocities.

In the atom, however, we cannot proceed so simply, because the nucleus exerts an attractive force on the electron, varying from one instant to another as their separation varies. The electron will speed up while it approaches the nucleus and slow down while it recedes; thus, its wavelength will vary as it moves. Its energy will be partly kinetic and partly potential, and the division between the two energies will vary.

The pendulum and the harmonic oscillator undergo a similar division of total energy, as shown in Chapter II. At the ends of its swing, a pendulum's energy is all potential; in the middle, its energy is all kinetic; and in between, its energy is a mixture of both. Figure 1 shows how this division of its energy varies with the instantaneous position of the pendulum.

The corresponding behavior of an electronic bit of stuff in an atom would be that pictured in Fig. 2. Circulating in an elliptical orbit with the nucleus at one focus of the ellipse, like a planet circulating about the sun, the electron would interchange potential energy (Fig. 2a) and kinetic energy (Fig. 2b) as it coursed its path.

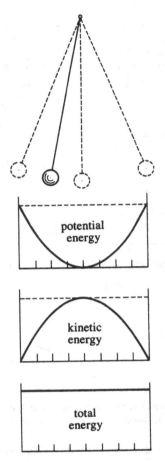

FIG. 1—A pendulum exchanging potential and kinetic energy as it swings.

And the sum of the two kinds of energy would stay constant as long as the electron did not gain or lose any of its energy by radiating it or by colliding with something else.

Hence, we cannot use de Broglie's relation to get information about the permitted energies in the simple way used before for the particle in a circular tube. But wave mechanics provides mathematical apparatus for calculating the shapes of the appropriate

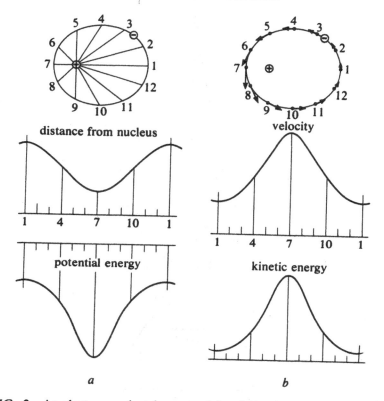

distance from nucleus

velocity

potential energy

kinetic energy

a   *b*

FIG. 2 — An electron exchanging potential and kinetic energy as it moves around the nucleus of an atom in an elliptical orbit.

waves. The waves have a length that varies from point to point. By that variation the sum of the kinetic energy (determined by the wavelength) and the potential energy (determined by how far the electron is from the nucleus) stays constant and so preserves a fixed total energy.

Before looking at the resulting probability smears that electrons can have in atoms, examine the shapes of the waves that wave mechanics would ascribe to a pendulum. Remember that the quantum theory permits only certain amounts of total energy to the pendulum, increasing in increments of $h\nu$, where ν is the frequency of the pendulum. To correspond with each of those permitted energies, wave mechanics provides a permitted waveshape (Fig. 3).

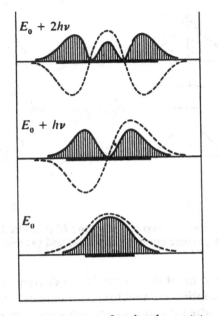

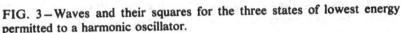

FIG. 3—Waves and their squares for the three states of lowest energy permitted to a harmonic oscillator.

Atomic States

Since the waves for an electron in an atom extend through three dimensions, they can be diagrammed only by taking cross sections. Figure 4 shows how such cross sections of two of the waves permitted to the electron in a hydrogen atom (1) roughly correspond with the two waves (3) for the pendulum and also (2) with two waves for a particle rigidly confined to move in one dimension over a limited distance.

The fact that all the waves for an electron in an atom tail off indefinitely shows that the electron can roam throughout space but is most likely to be found near the nucleus. In fact, an electron whose wave is either of those shown at the top of Fig. 4 is more likely to be found right at the nucleus than anywhere else.

The energy levels corresponding with the waves available to the electron in a hydrogen atom turn out to be arranged as in Fig. 5a. In contrast with the energy levels for the pendulum (shown in Chapter I), and for the particle on a circle (Fig. 11 of the last

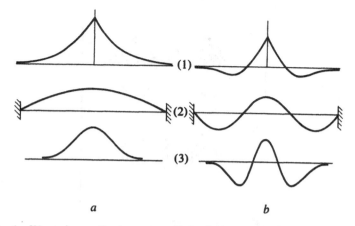

FIG. 4—Waveshapes for two states (*a* and *b*) permitted to (1) an electron in an atom, (2) a particle confined to a line, and (3) a harmonic oscillator.

chapter), the spacing of the atomic levels decreases with increasing energy, and a continuous spectrum of energies becomes available to the electron when it acquires enough energy to escape from the atom. In agreement with the customary convention, Fig. 5*a* chooses for the zero of energy the state in which the electron is at rest at a great distance from the nucleus.

Even in the earlier theory of the hydrogen atom—the theory of Rutherford and Bohr described in Chapter VI—only certain definite energy levels were permitted to the electrons. In that theory an electron with one or another of those energies could traverse one of only a limited collection of possible orbits, in which it would have the appropriate energy. Only one permitted orbit would give an electron the lowest permitted energy; there were four permitted orbits corresponding with the next higher permitted energy, and so on.

Wave mechanics replaced the idea of permitted orbits with the idea of permitted waves and therefore of permitted probability distributions. These permitted waves are the permitted stationary states, each associated with a definite permitted energy. In other words, the permitted stationary states are correlated with permitted energies, and usually there are several alternative states in which

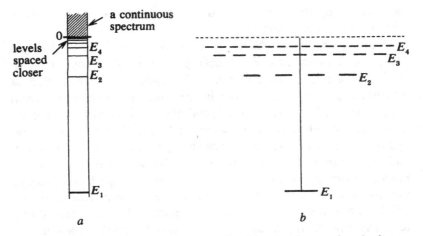

FIG. 5—(a) Levels of energy E_n permitted to the electron in a hydrogen atom. (b) The corresponding n^2 permitted states for the electron.

an electron would have the same energy, just as there were several orbits in the older theory.

For hydrogen, with only one electron, the permitted states and their energies can be calculated exactly. Figure 5b shows the same energy levels as Fig. 5a, with a number of lines n^2 at each level equal to the number of independent states in which an electron would have the energy E_n of that level.

Many-Electron Atoms

In any atom more complicated than hydrogen and therefore having more than one electron, the determination of permitted states and energies is likewise more complicated because the negative charges on the electrons make them all repel one another. But those states and energies can still be discussed by taking two things into account. In the first place, the nucleus of any other atom has a larger positive charge than the proton, the nucleus of hydrogen. Therefore, it will exert a larger attraction on every electron in the atom. But, second, that attraction will be opposed by the repulsions from any other electrons that are closer to the nucleus. In other words, any one electron is partly shielded from the nucleus by the opposite charge offered by the cloud of electrons between them (Fig. 6a).

Thinking about the effects of these two new factors—a larger charge on the nucleus and a shielding smear of electrons surrounding it—we can guess roughly what will happen to the states that an electron is permitted to occupy in the atom. For example, a state whose probability peaks up around the nucleus will peak even more sharply when the nucleus has a larger charge because that charge will pull the electron more strongly toward it. The electron will have less average potential energy because it is closer to the nucleus more of the time. It will have more average kinetic energy because it is more closely confined. But the gain in kinetic energy will be more than offset by the loss of potential energy, and, therefore, the total energy of an electron in that state will be lower.

A state whose probability is a maximum some little distance from the nucleus would be affected just as strongly by the increase in the nuclear charge if there were no other electrons in the atom.

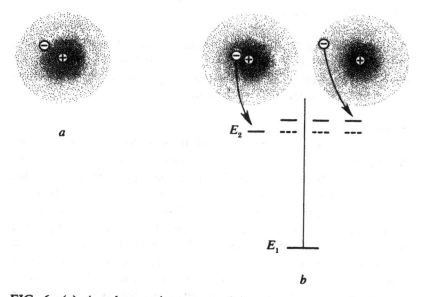

FIG. 6—(a) An electron in a many-electron atom shielded from the nucleus by its fellows. (b) Comparison of energies of (left) a less shielded and (right) a more shielded electron.

But when other electrons are present, those that spend time near the nucleus will partly shield the nuclear charge from an electron that spends time farther away. Then the total energy of the state will not be lowered so much by the increase in the nuclear charge. The contrasting situations are diagrammed in Fig. 6b.

The result of these two influences on the energies belonging to the states permitted to any one electron in an atom containing many electrons is shown in Fig. 7. Some of the states that correspond with the same energy in the one-electron atom (Fig. 5b) no longer have the same energy in a many-electron atom.

Interpreting the behavior of the electrons in atoms in this way, our first expectation would be that all the electrons would be found in the same state—the state with the lowest total energy. Any mechanical system, given a chance, will find a way to lose energy to its surroundings and fall into a condition in which it retains the lowest energy permitted to it. Its surroundings may raise its energy again if they have enough to do so and have some means of access by which they can deliver the energy. But the lowest energy permitted to the electrons in an atom is much lower than the next higher one permitted. The atom's surroundings are ordinarily too poorly endowed with energy to deliver the required amount to an appreciable number of electrons.

It is unexpected, therefore, to find that not all the electrons are in the state that would give them the lowest energy. Experiments suggest that at most two electrons are ever found at one time in

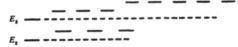

E_1 —

FIG. 7—The energies of some states permitted to an electron in a many-electron atom.

any one of the states permitted to them in an atom. In a hydrogen atom the single electron and in a helium atom the pair of electrons adopt that state. But in lithium and all the rest of the elements the additional electrons are forced to adopt states of higher energy.

Spin and Exclusion

The search for a rule to fit this curious fact led Pauli to propose the *exclusion principle* for electrons. That rule turns out to fit their behavior not only in isolated atoms but in molecules and solids as well, as you will see in the next chapter.

Shortly before Pauli made this proposal, Uhlenbeck and Goudsmit, studying spectroscopic observations of atoms in strong magnetic fields, had concluded that each electron must possess not only an electrical charge but also a magnetic dipole—that an electron is a little like a bar magnet. The required magnetic dipole could be produced by a constant electric current around a circular path.

It is important not to confuse this magnetic dipole with the magnetic effect that would result from the motion of the electron in an orbit around the nucleus. To be sure, that motion does produce a magnetic effect, but in addition the electron carries a magnetic dipole around with it (Fig. 8). Since the electron bears a negative charge and the magnetic effect could be produced by the motion of such a charge around a circle, the electron is sometimes visualized as a tiny ball of charge, spinning about an axis (Fig. 9c). This visualization has led to the name *spin* for the observed magnetic property of the electron.

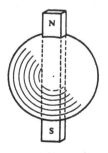

FIG. 8—An electron compared to a tiny bar magnet.

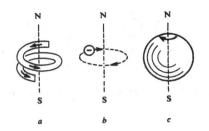

FIG. 9 – (a) A current-carrying coil of wire, (b) a charged particle traversing a circular orbit, and (c) a spinning ball of charge.

From the spectroscopic observations, Uhlenbeck and Goudsmit concluded also that the spin of an electron in a magnetic field imposed from outside could take only two possible values. The behavior of the electron suggested that its magnetic dipole always has the same strength and therefore that it always spins at the same speed. But it can spin in either a clockwise or a counterclockwise direction.

The restriction of the spin to two values is a little like the restrictions that quantum theory and wave mechanics put on the permitted states and energies of the electron. Using the language of those theories, you can say that the spin of an electron is apparently so strongly quantized in the presence of a magnetic field that it can take only two values. They are sometimes called *up* and *down*.

Now evidently whenever two electrons get near each other, they both find themselves in a magnetic field whether or not there is any magnetic field imposed from outside. Each is in a magnetic field caused by the spin of the other; each little bar magnet produces a magnetic field that acts upon the other little bar magnet. Hence, a pair of electrons close to each other can arrange their spins in one of only two ways: in the same direction or in opposite directions.

Pauli's proposal was that, when the two electrons are in the same stationary state – the same orbit in the older view – their spins can take only one of these two possible arrangements. According to his rule, their spins cannot be in the same direction but only in opposite directions (Fig. 10). Two spins in the same direction and in the same state are excluded.

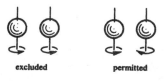

excluded   permitted

FIG. 10—Excluded and permitted spins of two electrons in the same state.

Shell Filling

Look now at some far-reaching consequences of the exclusion principle that help to explain the behavior of electrons in atoms. One permitted state in an atom gives an electron the lowest energy that it can have in the atom. The one electron in a hydrogen atom will normally be found in that state. The electron can be excited by enough energy from outside into a permitted state with higher energy. But after it has been excited, it will quickly find some way to deliver the extra energy back to its environment and so to return to the permitted state with the lowest energy.

In a helium atom, with two electrons, both the electrons will normally again be found in the permitted state of lowest energy. Since they are occupying the same state, those electrons will arrange their spins in opposite directions—one up and one down—as the exclusion principle describes (Fig. 11).

In a lithium atom, with three electrons, the third electron would again occupy the state of lowest energy if energy were the only factor controlling its choice. But if the third electron went into that state, its spin would have to be either up or down, and there are two electrons already present, one with spin up and the other with spin down. Hence, the third electron occupies the state next higher in energy (Fig. 12a).

Notice that lithium starts a new row in the Periodic Table. The

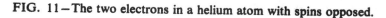

FIG. 11—The two electrons in a helium atom with spins opposed.

FIG. 12—(*a*) The three electrons in a lithium atom. (*b*) The five states of lower energy in a neon atom.

state into which its third electron is forced is one of the four in the next shell of states. Each successive species of atoms in that row in the table has one more electron. In neon all four states in that shell are occupied by electrons, and each of the four states has two electrons with their spins opposed (Fig. 12*b*). In sodium the additional electron is forced into a state in the next shell, and thus sodium becomes the first element in the next period of the table.

If the energies of the states in all the atoms were arranged as in Fig. 5*b*, the period beginning with sodium would contain 18 elements. The period would correspond with the filling of a shell of nine states, in which each state would accommodate two electrons. In fact, however, the electrons that fill the inner shells shield the outer electrons in five of those nine states more effectively from the nuclear charge. After eight electrons have completed the occupancy of four of the nine states, the energy of the next electron will be lower if it occupies a permitted state somewhat similar to the state occupied by the outermost electron in lithium or in sodium.

That electron makes potassium another alkali metal and places it in Group I of the Periodic Table, at the beginning of a new period. Another electron joins it, with opposite spin, in calcium. From this point, further electrons begin to occupy the five states previously neglected, and the first transition period in the Periodic Table arises.

In Fig. 13 the states for electrons in atoms are shown as boxes arranged on various levels. The higher the level of a box, the higher the energy of the state that the box represents. The levels do not

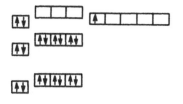

FIG. 13—Placing atomic states in rough order of energy to show why scandium begins a transition period between Groups II and III of the Periodic Table.

portray those energies to scale; they show only which energy is the larger.

Figure 13 can serve as a rough guide to the order in which the successive states are occupied by electrons in pairs to form the successive atomic species in the Periodic Table. But no single diagram of this sort can be a completely accurate guide. The positions of the levels really shift as the addition of the electrons and the corresponding increase in the nuclear charge progress, and sometimes the levels shift differently enough to change their vertical order.

Already, in the many-electron atom, we must abandon hope of making exact calculations of the shapes of the permitted electronic waves. The number of particles all interacting—the nucleus and the several electrons—is too great for the mathematical problem to be worked exactly. The probability waves for electrons in a solid, with Avogadro's number of atoms instead of only one, must be studied even less exactly. But you may be surprised to find, in the next chapter, how well the wave-mechanical principles described in the last two chapters can provide a picture of the behavior of the electrons in molecules and in solids.

XII. ELECTRONS IN SOLIDS

*In order to obtain a consistent account of atomic phenomena,
it was necessary to renounce even more the use of pictures.*
NIELS BOHR, *Atoms and Human Knowledge*

WHEN Schrödinger laid the basis for wave mechanics in 1925, he
used it to calculate the stationary states of the electron in a hydrogen atom and the energy belonging to each state. The calculated
energies checked almost exactly the information that had been
accumulated by experiments on hydrogen. His achievement made
it clear that wave mechanics might provide, for the first time in the
development of physics, a penetrating description of the electronic
behavior of matter.

Most of the immediately following work employed the new
theory to explain the behavior of isolated atoms. But within three
years Arnold Sommerfeld had applied the theory to solids. He
showed how Schrödinger's wave mechanics and Pauli's exclusion
principle could explain much of the behavior of the free electrons
in a metal. Simple though his method was, it showed convincingly
why those electrons do not contribute to the heat capacity of a
metal and so removed one of the inconsistencies mentioned at the
beginning of Chapter X.

Free Electrons in Metals

To make a simple wave-mechanical theory of the behavior of
electrons in a metal, Sommerfeld supposed that the electrons are
perfectly free to move around inside the metal but cannot get out.
The first part of this supposition—that the electrons are perfectly

free inside the metal—is an approximation, of course. In a real metal, however free the electrons may be, they are still influenced by the ions that they have left behind them. But the approximation makes a good starting place.

The second part of the supposition—that the electrons cannot get out of the metal—is much better. Actually, electrons normally do not escape from a metal except when it is very hot or when it forms part of an electric circuit in which there is an electric current, so that the electrons lost at one end are replenished from the other. But to say that the electrons cannot get out of the metal is the same as to say that each electron is confined. Hence, again a set of discrete stationary states will be permitted to an electron, as the last chapter described.

In studying the behavior of the particle in Fig. 10 of Chapter X—a particle that is free to move around a circle as long as it stays on the circle—you have already seen waves that are very similar to the waves for electrons that are free to move within the confines of a box. In fact, if the tube of very narrow bore used in Chapter X is not bent into a circle but left straight and blocked at both ends, the tube forms such a box. In other words the electrons are then confined to a straight line rather than a circle—a line having two ends beyond which the electrons cannot go (Fig. 1).

Thus, an electron wave will be a simple wave like that of an electron on a circle; but now the wave must go to zero at both ends of the line. The only waves that can do so are those with half-lengths that fit evenly into the length of the line. Some of the permitted waves are shown in Fig. 2.

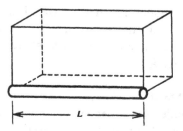

FIG. 1—A thin tube blocked at both ends to form the simplest box in which to imagine confining electrons.

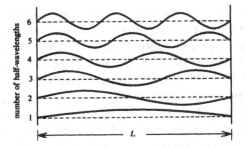

FIG. 2—The six waves of the states of lowest energy for an electron in the tube of Fig. 1.

Like the electrons on a circle, electrons free to move on a line have only kinetic energy, no potential energy. An argument the same as that used in Chapter X for the electrons on a circle gives the permitted values of energy for the waves. Very similar conclusions can be reached for electrons free to move in the larger box of Fig. 1, which better represents a piece of metal.

Pauli's exclusion principle now provides the guide to the way in which electrons will distribute themselves among these states. Since only two electrons, one with spin up and the other with spin down, can occupy any one of the states, the electrons will fill up the permitted states of lowest energy in pairs, just as they do in building atoms (Chapter XI), until all of them have found permitted states. But there are so many free electrons in a metal that some of them must occupy states of very high energy (Fig. 3).

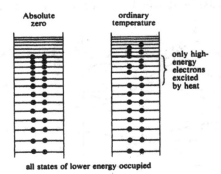

FIG. 3—Energies of electrons in a metal.

You might think at first that this argument depends on having a piece of metal large enough to include a very large number of electrons and would not apply to a microscopic particle of metallic dust. But the smaller the box, the larger the spacing between permitted levels (Fig. 4). Conversely, doubling the size of the crystal represented by this model cuts in half the spacing between the permitted levels and at the same time doubles the number of electrons that must be accommodated. Hence, in the state of highest energy that a pair of electrons will be forced to adopt, their energy will be the same whatever the length of the line to which they are confined if the total number of electrons present is proportional to the length of the line.

In the bulkier box of Fig. 1 it turns out that again the decrease in the number of electrons, which comes with decreasing the size of the box, is exactly compensated by the increase in the spacing of the permitted energy levels. For all boxes, whatever the size or shape, a few electrons must have the same very high energy. That energy depends only on the number of electrons per unit volume. Hence, in any metal that is well represented by this model, the energy of the most energetic electrons depends only on the nature of the material, not on the size of the sample.

Now imagine what happens when this electron gas is heated. Since heat is a form of energy, the electrons can absorb heat only by going into states of higher energy. And an electron with low energy cannot be excited to any state of higher energy that is already occupied by a pair of electrons.

In other words, heat will be absorbed by electrons only in the

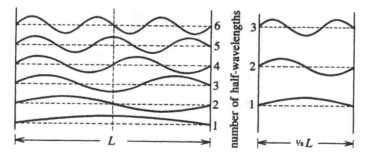

FIG. 4 — Cutting a piece of metal in half.

course of exciting them to the states that are not already fully occupied. But an electron can go only to one of the states of very high energy that are above the top of the occupied heap, and the heat energy is too small to make it do this unless it has quite a high energy already. The heat energy can excite only the few electrons that are in states whose energies are near the top of the heap. For this reason the free electrons in a metal can absorb only a negligible amount of heat; the electron gas makes practically no contribution to the heat capacity of a metal under ordinary circumstances. The atomic vibrations are the only contributors (Chapters II and III).

There are extraordinary circumstances in which the heat capacity of the electron gas might be detected; we can predict how to look for it by examining how it should vary with temperature. As the temperature increases, the heat energy in the surroundings also increases and can excite more electrons to higher levels of energy. It turns out that the heat capacity of the electron gas should be directly proportional to the temperature at which it is measured, following the sloping line in Fig. 5a.

At very low temperatures the heat capacity contributed by the vibrations of the atoms follows a very different curve (Fig. 5b), as Chapter III described. Hence, at those low temperatures a tiny heat capacity, increasing in proportion with the temperature, will not be masked by the atomic vibrations; it can be ascribed to the electron gas.

The heat capacity of the electron gas might be detected again at very high temperatures. There the heat capacity of the atomic vi-

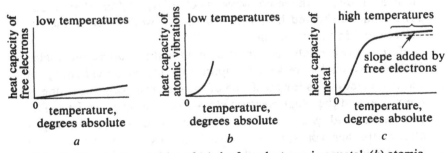

FIG. 5—The heat capacities of (a) the free electrons in a metal, (b) atomic vibrations at very low temperature, and (c) a metal at very high temperature.

brations should have leveled off at the Dulong-Petit value and should no longer vary with the temperature. Then the heat capacity of the electron gas, still increasing with the temperature, should add a small positive slope to the level line (Fig. 5c).

The Hydrogen Molecule

In order to make much more progress in picturing the behavior of electrons in a solid, we must take into account the presence of their parent atoms. In a metal the electrons do not inhabit an empty box but one populated also by the positive ions that the electrons leave behind when they stray from the atoms.

In order to form a model of a metal, even the empty box has, somewhat surreptitiously, made a crude use of those ions. The electrons cannot leave the box only because those ions are there. In the absence of the attractive force exerted on them by the positively charged ions, the negatively charged electrons would instantly fly away from one another in all directions.

But the ions have also a more detailed influence on the electrons. Since each ion attracts every electron, any electron is a little more likely to be found near an ion than farther away. This effect influences the forms of the waves for the electrons so that they look somewhat different from the simple waves for an electron in an empty box that were shown in Fig. 2.

A good way to begin examining this effect is to forget boxes for a moment and think about how electrons would behave in the simplest possible substance that contains more than one attractive ion. That substance is a molecule of hydrogen. Less than two years after Schrödinger invented wave mechanics, Walter Heitler and Fritz London showed how the new theory would picture the covalent bond in that molecule.

Here the two electrons circulate about just two attractive nuclei. If those two nuclei are very far apart, then the combined energy of the two electrons is lowest if one of the electrons stays near one of the nuclei and the other electron stays near the other nucleus. The two nuclei and two electrons will form two isolated hydrogen atoms. In other words, the two nuclei offer two favorable states to the two electrons, which are the states corresponding to the lowest energy in the two atoms.

When the two nuclei get closer together, they again offer two low-energy states to the electrons. But those states no longer correspond to quite the same energy. The difference in energy between those states grows larger as the two nuclei get closer to each other (Fig. 6). Both electrons adopt the state of lower energy, arranging their spins so that one is up and the other is down.

Hence, the two electrons can have a lower energy when the nuclei are close together than when they are farther apart. That fact is sufficient to insure that the nuclei will approach each other until the electrostatic repulsion between them increases enough to outweigh the decrease in energy of the electrons. In other words, lowering the electronic energy constitutes a bond between the two atoms.

By looking at the forms of the waves for electrons in these molecular states, you can connect the wave-mechanical picture of the bond in the hydrogen molecule with the earlier picture of the covalent bond in Chapter VII. When the nuclei are far apart, the two waves have the shape shown in Fig. 7a. When the nuclei are near each other, the two waves change into the shapes shown in Fig. 7b. The probability distributions for an electron, obtained by squaring those waves (Fig. 8), shows that there is an equal probability of finding the electron in either of the two atoms that form the molecule, no matter which state the electron is in.

The two states differ, however, in the probability that the electron is between the nuclei. In the lower energy state, the electron is more likely to be between the two nuclei than outside. In the

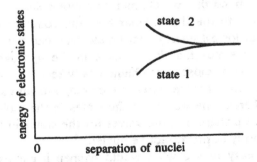

FIG. 6—The energies of the two states of lowest energy for electrons in a hydrogen molecule.

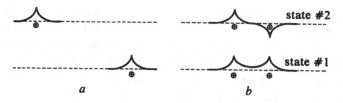

FIG. 7 – Electron waves of the two lowest energy states for electrons in a hydrogen molecule when the atoms are (a) far apart and (b) close together.

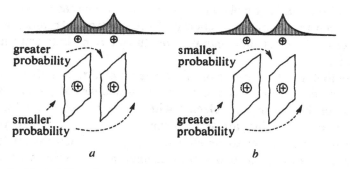

FIG. 8 – Probability distributions for an electron in two states of Fig. 7.

higher energy state, the probability is greater that the electron is outside than that it is between the nuclei. In other words, an electron tends to pull the nuclei together by its electrostatic attraction if it is in the lower energy state and to pull them apart if it is in the higher energy state. Both the electrons, by occupying the lower energy state, bond the two atoms into a molecule.

Return now to the thin tubular box and compare in Fig. 9 the two states of lowest energy that an electron can have in that box with the two states that it can have in the hydrogen molecule. There is a fairly close resemblance betweeen the two sorts of waves. The waves for the states in the box go to zero at the ends of the box, whereas the waves for the states in the molecule tail off indefinitely. Furthermore, the waves for the states in the molecule peak up sharply at the attractive nuclei.

But it is easy to see what would happen if enclosing a single hydrogen molecule in a very small box combined the two situations. Two attractive nuclei placed one-quarter of the way from

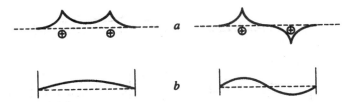

FIG. 9—Comparison of waves for an electron (a) in a hydrogen molecule and (b) in a box.

either end of a box will modify the two states of lowest energy in the box by making the waves for those states peak up at the nuclei (Fig. 10). In other words, there will be a greater probability that the electron is near one or the other of the nuclei, because they attract it. More generally, in the neighborhood of either nucleus the wave for a state will look like the wave in an atom and farther from a nucleus like the wave in a box.

Electron Waves in Crystals

These waves for states of an electron in a box containing two nuclei form a simple starting point for thinking of waves for states when the box contains more nuclei. When there are only two nuclei, the shapes of the two waves in the lower half of Fig. 10 can be drawn in the following way.

First plot a waveshape (Fig. 11a) that looks like the wave for the state of an electron in an atom. Repeat that shape at each of the atoms and connect those repetitions smoothly (Fig. 11b). Then, in order to find the two final waveshapes, multiply the repeated atomic waveshape by the shapes of the two waves (Fig. 11c) for

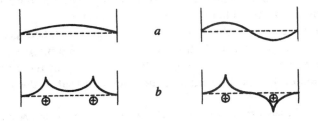

FIG. 10—Modification of waves for (a) an electron in a box by (b) adding two nuclei.

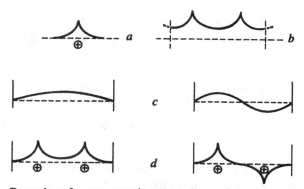

FIG. 11—Procedure for constructing waveshapes for an electron in a box containing attractive nuclei.

the empty box. In other words, at each point along the horizontal line, multiply the height of the wave in Fig. 11*b* by the height of one of the waves in Fig. 11*c*. The final waveshapes (Fig. 11*d*) are those already shown in Fig. 10.

Two effects of that procedure of multiplication are conspicuous. The wave of lower energy in the empty box (Fig. 11*c*, left) has the same value at the positions of the two nuclei. Therefore, the product wave is symmetrical and looks the same at both nuclei. The wave of higher energy in the empty box (Fig. 11*c*, right) reverses its sign at the middle of the box. Therefore, the sign of the product wave at one nucleus is opposite to its sign at the other nucleus, and the wave goes to zero midway between the nuclei.

The same procedure can be used to picture the waves in a box containing more nuclei, evenly spaced as they are in a crystal. Hardly had Sommerfeld developed his theory of the electron gas when Felix Bloch refined it in this way to provide a more realistic description of the states of electrons in crystals.

Figure 12 shows the result of applying that procedure to the waves for a particle in a thin tubular box containing six evenly spaced nuclei.* The six waves of lowest energy in the empty box are modified by the presence of the nuclei so that in the neighbor-

---

*For valid quantitative calculations using waves obtained by this procedure, the form of the repeated waveshape at the top of the figure must be changed slightly and differently for each of the waves in the empty box before multiplying the two. But the picture remains qualitatively unchanged.

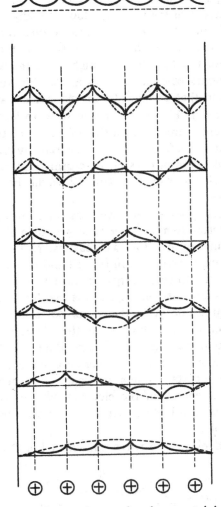

FIG. 12—Waves for an electron in a box containing six nuclei.

hood of those nuclei they look like the wave for an electron in an atom. And at the same time the probability that an electron in one of these states would be found near any particular nucleus is modified by one of the waves appropriate to the empty box.

There is nothing remarkable in the fact that the closer an electron gets to a particular ion, the more its behavior is dominated by that ion. Its wave near the ion might look like any of the waves that it could have in the atom. In Fig. 12 the waves are drawn as if the electron when near an ion had a wave such as Fig. 4a in the last chapter showed. But the electron might have the wave pictured in part b of that figure when it was near an ion. Figure 13 shows what would happen in that event. The lowest energy wave of Fig. 12, shown again in Fig. 13a, would be changed to look like Fig. 13b. The two different atomic waveforms of Fig. 4 of the last chapter, shown again at the left of Fig. 13, generate by the procedure of Fig. 11 the two different crystal waveforms that are shown at the right. Indeed, all the waves in Fig. 12 would be changed analogously.

Electrons having waves that were modified in this second way would almost certainly have higher energies than electrons with the waves shown in Fig. 12, for two reasons. In the first place, the shape of the atomic wave used in Fig. 12 and Fig. 13a is that for an electron having the lowest energy permitted in the atom, whereas the atomic waveform used in Fig. 13b is that for an electron having a higher permitted energy in the atom. An electron has a higher energy near each atom in the crystal, therefore, if its waveform in that neighborhood has the latter shape than if it has the former shape. Since the states of interest here are stationary states for an electron in the crystal—states in which its energy stays constant—the higher energy of the electron when it is near an atom must be reflected in a higher energy for the stationary state in the crystal as a whole.

In the second place, the wave shown in Fig. 13 has more ups

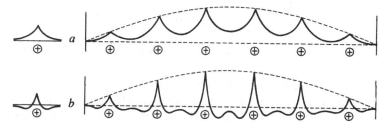

FIG. 13—Waves for different electronic states in an atom constructing different waves in a crystal.

and downs in it than the corresponding wave in Fig. 12. A wave that oscillates up and down over a shorter distance has a shorter de Broglie wavelength than a wave with fewer oscillations. We have seen before that the shorter the wavelength for a particle, the higher will be its velocity and therefore also its kinetic energy. Again for that reason, the stationary states represented by the second batch of six waves should correspond with higher energies than the states represented by the first batch.

Notice that each of these batches contains six permitted states. This is not an accident; if there are six ions present, they will offer six states for each state that the electron could have on one of the isolated atoms. When the six ions are very far apart, for example, they offer batches of six states. One of the states in each batch is a state for an electron on one of the isolated atoms. As long as the atoms are far apart, the six states in any one batch all give an electron the same energy. But they are not all the same state, because each belongs to a different atom.

When the atoms get close together, so that an electron can move from one to another, the energies appropriate to the six states begin to differ from one another. An interesting analogy connects the resulting behavior of the electrons with the coupled oscillators in Chapter III. The six states are analogous to the six oscillators of Fig. 9 at the end of Chapter III, and the electronic energies are analogous to the vibration frequencies of those oscillators. When the oscillators are uncoupled, they all vibrate at the same single frequency, just as the electronic states on all the atoms have the same energy when the atoms are far apart. When the six oscillators are coupled, they can oscillate collectively at six different frequencies; when the six atoms are near one another, they collectively afford six states with differing energies for the electrons.

Bands of Permitted Energies

These states are, however, even more strongly reminiscent of the states for electrons in the hydrogen molecule. A diagram of the permitted energies of an electron in a crystal (Fig. 14) is analogous to Fig. 8 drawn for the permitted energies of an electron in the hydrogen molecule. The new diagram takes into account the fact that there are six atoms in the crystal instead of the two in the

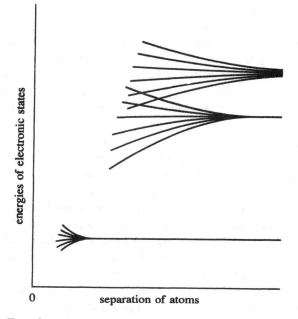

FIG. 14—Energies of states for an electron in a crystal separating as the atoms are brought closer together.

molecule and also the fact that each isolated atom offers more states than one. For the hydrogen molecule Fig. 8 took into consideration only one of the states offered by each hydrogen atom — the state of lowest energy.

The analogy with the hydrogen molecule makes it plausible to draw such a diagram as Fig. 15, plotting the energies permitted to an electron when a great many atoms are brought together in a regular arrangement with a fixed known spacing between the atoms. The permitted energies of the states that an electron could occupy on the isolated atoms are split into batches of permitted energies for an electron in the collection of atoms. Each batch of permitted energies is called an *energy band*.

Each of these energy bands corresponds to just one of the energies permitted in an isolated parent atom. If there are N atoms in the crystal, there are N energies in each band. But the upper and lower extremes of each band depend only on what elements make

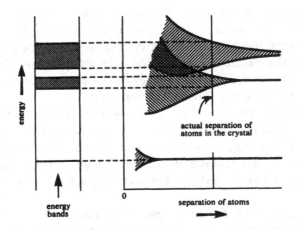

energy →

energy bands

0   separation of atoms →

actual separation of atoms in the crystal

FIG. 15 – Energy bands formed by the energies permitted to an electron in a crystal.

up the crystal and on the arrangement of the atoms, and not on the size of the crystal.

The fact that the band edges do not depend on the size of the crystal is much like the conclusion, reached earlier in Fig. 4, that the maximum energy of a free electron in a metallic crystal is independent of the size of the crystal. For the band edges this conclusion follows upon comparison (Fig. 16) of the waveforms for an electron with the lowest and highest energies with the two waveforms for an electron in the hydrogen molecule confined to a box.

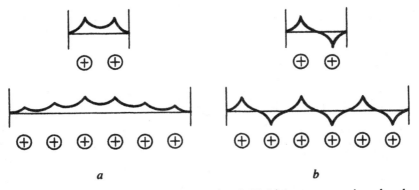

a   b

FIG. 16 – States of (a) lowest energy and (b) highest energy in a band.

An electron in the state of lowest energy, for example, has a waveform very similar to that in the bonding state of the molecule. In the crystal the only important difference is that the probability of finding the electron is distributed over more atoms. The waveform for the state of highest energy in the crystal is analogous to the form for the other state in the molecule.

Thinking of the molecule as a crystal with only two atoms, you see that increasing the size of the crystal increases only the number of states with energies that fall between the bounding energies. The bounding energies themselves do not change with the size of the crystal so long as the distance between the neighboring atoms stays the same.

Calculating from first principles the location of the bounding energies of the bands for a particular crystal is a difficult task. The calculation has been done approximately for a few materials. When such an approximation is available, its accuracy can often be improved by making various sorts of experimental measurements on the material.

Even without quantitative results, however, the idea of energy bands can be used to gain insight into why some materials are electrical conductors and others are insulators. You will get that insight in the next chapter by examining how the electrons occupy the states corresponding to the energies in the permitted bands and how they respond to electrical forces that try to move them.

XIII. ELECTRICAL CONDUCTION

And when the rain has wetted the kite and twine, so that it can conduct the electric fire freely, you will find it stream out plentifully from the key on the approach of your knuckle.
BENJAMIN FRANKLIN, *Letter to Peter Collinson*

IN PICTURING the bonding of atoms into solids, Chapter VII made clear that metals conduct electricity well because they contain free electrons and that insulators insulate because their electrons cannot readily move out of the ions or out of the bonds that form them. But these pictures leave many questions unanswered. Why do some substances have free electrons, while others do not? Why does the large current that a metal will carry decrease when the temperature increases, while the small current that an insulator will carry increases with the temperature?

At the outset, answering the first question seems fairly straightforward. Any atom will free an electron if enough electrical force is exerted on it to pull the electron out. When the atoms get close to one another in a liquid or a solid, the electrons in each atom are attracted by the positive charges on the nuclei of the neighboring atoms. If those forces are strong enough, an electron leaves the atom.

The best way to put numbers into that answer is to speak of energies rather than forces. The amount of energy required to pull an electron out of an atom would be the amount required to lift an electron from the energy level that it occupies in an atom—some level like those shown in Fig. 5 of Chapter XI—up to the zero level, where it is free.

From experiments on isolated atoms, those energies have been measured and tabulated. Figure 1 shows the least of those energies for each element—the *first ionization potential*—the energy required to remove one electron from the topmost occupied energy level in an atom. The unit of energy used in that diagram is the electron-volt, which is described in the Appendix. Proceeding through the diagram and identifying the elements that are metals and those that are not, you will find that there is a fairly definite dividing line running horizontally through the diagram. In other words, at the atomic separations found in solids, the neighboring nuclei seem to be able to exert about the same amount of force on one another's electrons no matter what element they are. If the electrons need more force to escape, they will not be free to conduct electricity.

This argument says little about whether a liquid metal should show the same behavior as a solid metal. It says only that any difference between them might be ascribed to the slightly greater separation of the atoms in the liquid. As a matter of fact, liquid metals are very good conductors of electricity. The strongest hint of trouble in explaining conduction in liquid metals is the fact that the resistance to the flow of electricity in most metals nearly doubles when they melt. That seems a large change to explain by the very small increase in the atomic separation.

It turns out that these questions have been answered only by the wave interpretation of electronic behavior, as pursued in the last few chapters. But before resorting again to that interpretation for help, it is worth while looking at what the picture of electrons as bits of stuff can do to make clear the process of electrical conduction.

Conductance and Resistance

An electrical force applied to an electrified "bit" will start it to move; if the force continues to act on it, the bit will gain speed. Now the motion of an electrical charge constitutes an electric current; the amount of current is given by the amount of charge times the velocity at which the charge moves. Hence, the current carried by the electrified *bit* will keep increasing as the bit gains

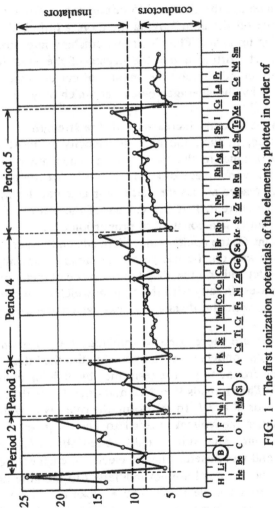

FIG. 1 – The first ionization potentials of the elements, plotted in order of their listing in the Periodic Table.

speed until finally the bit will crash into something and the current will stop abruptly.

When the electrified bit is a free electron in a metal, it will crash into an atom before the electrical force can accelerate it very much. It will bounce off the atom, giving the atom a little kick, and then start speeding up again. The electrons, crashing into atoms whenever they start drifting under the urging of the electrical force, cannot speed up indefinitely; the amount of current that they carry is determined by the average velocity at which they all succeed in drifting.

That average velocity increases with the strength of the applied electrical force; in fact, the average velocity in most metals is directly proportional to the force. As an immediate consequence, the current through a piece of such a metal is directly proportional to the voltage placed across the piece, or in algebraic form, $I = GV$, where I is the current, V is the voltage of the source of the electrical force, and G is a constant of proportionality.

The constant G, the *conductance*, depends on the size and shape of the sample and on the nature of the metal composing it. The conductance depends on the shape of the piece of metal in a simple way: It increases directly with the area on which the force is acting and decreases with the distance across which the voltage is applied.

Most people are more used to thinking of the *resistance* of a metal than of its conductance, and you can turn the little rule of the last paragraph around to get the resistance by dividing both sides of the rule by G. Then it reads $V = I/G$, and since $1/G$ is the resistance R, the rule reads $V = RI$. In this form, the rule says that the larger R is, the higher the voltage needed to keep a given amount of electricity flowing. That rule, usually called *Ohm's law*, is one of the simplest and most useful rules of electrical engineering.

The atoms of the metal are not indifferent to the little kicks that they receive from the electrons. Those kicks increase the amplitudes of the atomic vibrations; the temperature of the entire piece of metal rises as energy is transferred to the atoms by the collisions of the electrons. The rate at which the energy is transferred increases with the velocities that the electrons have when they collide and also with the force that is pushing them. In consequence, the rate at which the metal becomes hotter is proportional to the product of the voltage V placed across the piece and the current I through it.

Variation of Resistance with Temperature

Now look a little more closely at what the free electrons must be doing in the metal. If they truly form a gas of free electrons, they must be dashing about as do the atoms in the more familiar gases. In fact, many of them must have extremely high speeds. The last chapter showed that many of them are in states where their ener-

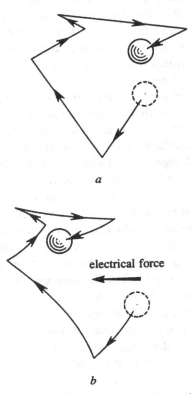

a

electrical force

b

FIG. 2—(*a*) A free electron tracing an irregular path at high speed in a crystal. (*b*) An applied electrical force adding a slight directed drift.

gies are very great (Fig. 3 of Chapter XII) and also that those energies must be largely kinetic because the wavelengths of the high-energy electrons are very short (Fig. 2 of Chapter XII).

The behavior of those electrons, therefore, must be somewhat as pictured in Fig. 2. They thread their way among the atoms rapidly and irregularly whether or not any electrical force is applied to them, bouncing off the atoms and gaining or losing energy on many of those bounces. On the average, they neither gain nor lose, because the electron gas and the atomic grid are in thermal equilibrium with each other.

An applied electrical force adds a slight directed drift to the paths of the electrons. But that drift is small in proportion to the random velocities that the electrons already have—far too small to make them collide with the atoms more frequently. The only effect of the drift on the collisions is to transfer some of the changed velocities to the atoms because the drift is a departure from the condition of thermal equilibrium.

Now try to fit into this picture of electrical conduction the experimental observation that the resistance of almost all metals over a very wide range of temperatures increases in direct proportion to the absolute temperature. To see what the picture would predict, turn first to the change in the behavior of the grid of atoms that a higher temperature will make. The agitation of the atoms of the metal will increase, as Chapters II and III described. But the higher velocities of the atoms cannot have much effect on the rate at which the electrons collide with them, because the velocities of the atoms are much less than the velocities of the electrons.

Turn then to the gas of free electrons. If that gas behaved like a gas of atoms, an increase in temperature would increase the velocities of the electrons. The electrons would collide more often with the atoms and would have less time between collisions to acquire a drift from the electrical force. Hence, their average drift velocity, on which the current depends, would be less. The current would be less for the same electrical force; that is, the resistance would be increased.

But two facts already discussed make this argument unsatisfactory. In the first place, Chapter II, discussing the heat capacity of an atomic gas, pointed out that the average kinetic energy of the atoms in the gas increases in direct proportion to the temperature. But in order to use the present argument to explain the fact that the resistance of a metal increases directly with the temperature, the average random speed of the electrons—not their average energy—must increase directly with the temperature. Since the kinetic energy is proportional to the square of the speed, the present argument suggests that the resistance should increase with the square root of the temperature, and experiment does not bear out that suggestion.

In the second place, Chapter XII pointed out that the electron gas in a metal does not behave like a gas of atoms; its heat capacity is negligibly small. It does not acquire much additional kinetic energy when the temperature rises, because, as Fig. 3 of Chapter XII showed, most of the electrons cannot accept any energy. But if a rise in temperature cannot greatly change the speeds of the electrons, it is hard to see how a change of temperature can greatly change their rates of collision with the atoms. By this argument, the resistance should not vary at all with the temperature.

Scattering Electron Waves

These difficulties can be resolved by taking more fully into account the wave nature of the electron. In fact, we can explain in that way not only the variation in the resistance of metals with temperature but also the distinction between metals and insulators.

Here again, as in the last chapter, it is easiest to picture the behavior of the electron waves by dealing with a tube of narrow bore, blocked at both ends, that contains a row of positively charged nuclei regularly spaced and an appropriate number of electrons.

Certainly the electrons must be moving in that tube because their wavelengths cannot be infinitely long and, hence, by the de Broglie equation their velocities must be something other than zero. There are two directions in which any particular electron can be moving: to the right or to the left.

A single diagram (Fig. 3) can display both the velocities and the energies permitted to the electrons in the tube. First suppose that the tube contains no nuclei—that the electrons are truly free. Then the lengths of the permitted simple waves, already shown in Fig. 2

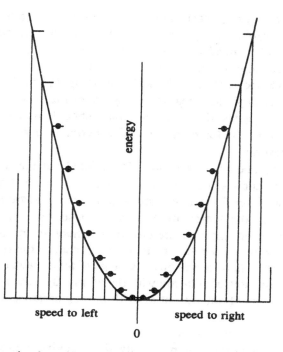

FIG. 3 — Permitted speeds and energies of an electron free in a tube of narrow bore blocked at the ends.

of Chapter XII, will determine the permitted velocities, evenly spaced. They are plotted horizontally in Fig. 3. Corresponding to each of the permitted velocities is a permitted energy, plotted vertically in Fig. 3. Since the tube contains no nuclei, the energy is all kinetic and is therefore proportional to the square of the velocity. The same number of electrons are traveling in each direction; there is no net electric current.

Now suppose that an electrical force is applied to the electrons, urging them to move toward the right. It will add to the speeds of the electrons that are already moving to the right and subtract from the speeds of those moving to the left, more and more as time passes (Fig. 4).

Finally imagine that nuclei are placed in the tube, regularly spaced along it. Evidently, the process pictured in Fig. 4 could no longer continue indefinitely. By adopting states further and further

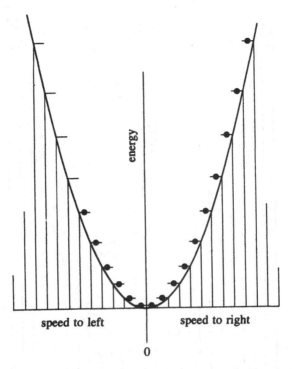

FIG. 4—An electrical force moving free electrons into states of higher velocity in its direction.

to the right side of Fig. 4, the electrons will finally reach the upper edge of the band of permitted energies. In order to show this, Fig. 3 needs to be modified to look like Fig. 5.

What will happen when an electron finally reaches a velocity that brings it to a band edge? That question can be answered through a variety of approaches which all lead to the same conclusion. One approach, available from what this book has already discussed, is to recall (Chapter X) that electrons can be diffracted by crystals. The conditions for their diffraction are the same as those for the diffraction of X rays by crystals, described in Chapter IV. In particular, an electronic wave will be strongly reflected when its wavelength and its angle of approach to the planes of atoms in the crystal have a suitable relationship to the spacing between those planes.

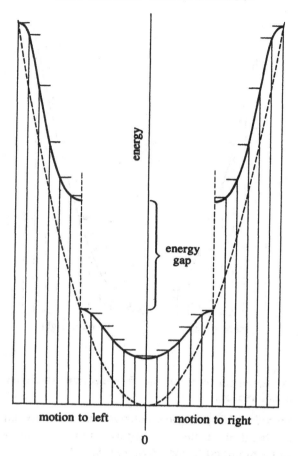

FIG. 5—Two of the bands of speeds and energies permitted to an electron in a tube containing nuclei.

That relationship is exactly fulfilled by the wavelength of the electron in the state whose energy is at the upper edge of a permitted band (Fig. 6). In order to verify that Bragg's law (Chapter IV) is satisfied here, notice that the wavelength λ of the electron is twice the atomic spacing d and that the glancing angle is 90 degrees. Hence, an electron reaching the state at the band edge on the right of Fig. 6 will be reflected and go to the left (Fig. 7).

Now consider the effect of temperature on such processes. Since the thermal agitation of the atoms in the solid is a chaotic motion,

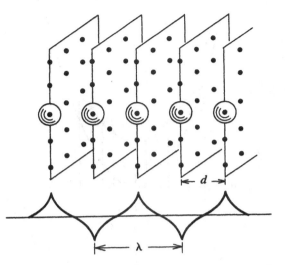

FIG. 6 – Bragg reflection from the nuclei when the wavelength of an electron is twice the internuclear distance.

the solid presents to the electrons at any instant a somewhat disordered arrangement of atoms. Even though the arrangement is orderly on the average, the atoms are not all evenly spaced at any one time. Electrons of wavelengths other than the wavelength at the band edge can be reflected also because they encounter atomic spacings other than the average spacing. In this way the electrons are scattered backward long before they reach the state at the band edge.

Figure 8 presents a simplified diagram of the conduction process just described. Electrons are accelerated to the right and are scattered back to the left. They find at the left side of Fig. 8 states whose energies are lower than the energies of the states that they deserted on the right side of Fig. 8. The energy that they give up in passing from right to left is accepted in the form of heat by the atoms that are reflecting them.

It is certainly clear that the resistance will increase with increasing temperature. The instantaneous disorder of the atoms is greater at high temperatures, and they can scatter waves of more diverse wavelengths more effectively. But it is not easy to see why the resistance should increase in direct proportion to the temperature;

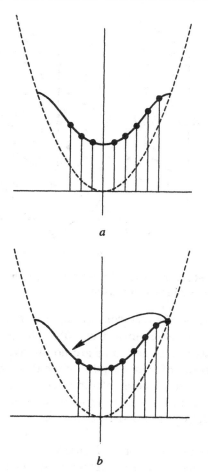

FIG. 7—An electron accelerated (*a*) to the right is reflected (*b*) to the left when it reaches the band edge.

the wave-mechanical theory deducing that dependence is too complicated for this book.

Wave Scattering and Collisions

The picture of electrons as waves, scattered increasingly with increasing atomic disorder, explains why liquid metals have electrical resistances so much higher than solid metals. And it explains

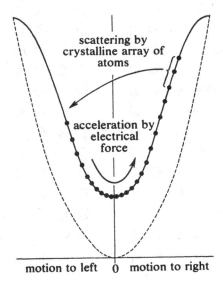

motion to left   0   motion to right

FIG. 8 — Electron waves scattered by the instantaneous atomic disorder due to thermal agitation.

the interesting fact that alloys of two metals almost always have higher resistances than either of the metals alone. The different size of the atoms of the alloying metal warps the arrangement of atoms in the alloyed metal to provide a range of atomic spacings that increases the scattering power of the array.

Carrying these ideas of electronic scattering from a tube of narrow bore into a three-dimensional solid, we can readily relate them to the earlier picture of the collisions of particles. Each electron is a burst of waves extending over many interatomic distances and traveling in some direction. The burst advances until it meets a portion of the crystal in which the atomic spacings are momentarily suitable to reflect the wave at one or another angle, not necessarily straight back. Then the burst of waves is deflected by that portion of the crystal from its original path. The picture becomes a transcription of collisions in terms of waves (Fig. 9).

There are three major differences between the two pictures of collisions. In the first place, a collision of the waves cannot be localized to a single atom but only to a region containing an array of atoms. Moreover, a collision of the waves is described by the

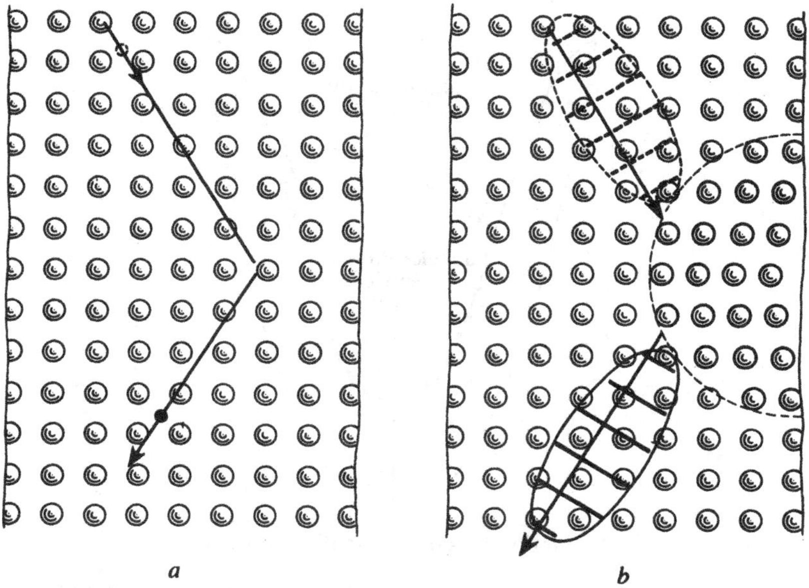

FIG. 9 — Collisions of atoms in a crystal with an electron regarded (*a*) as a particle and (*b*) as a burst of waves.

rules of diffraction of waves rather than by the rules of colliding billiard balls. And, finally, the waves, whose lengths are so vital in the new rules of collision, have properties described by the exclusion principle when the colliding waves are electronic.

Conductors and Insulators

Examining that last fact again, we can quickly reach a conclusion that explains the conspicuous differences between electrical conductors and electrical insulators. In the argument about the reflection of electronic waves in the tube of narrow bore, the electrons could always be accelerated when an electrical force was applied. In other words, they could always find vacant states to the right in Fig. 7, into which they could move until they reached the band edge. What would happen if they could not find such states?

Then the material would be an electrical insulator. If all the states in one of the bands pictured in Fig. 5 were already occupied by electrons, an electron could not advance to the right into one of those states. A way might be made for the electrons to move to the right by reflecting an electron at the right edge of the band back to the left edge of the band. But that would still provide no electric current, because the total velocity of the entire collection of electrons would still be zero (Fig. 10). Moreover, the electron at the band edge could not be reflected, because it would find itself going in the other direction with the same wavelength and would be reflected back again.

In other words, an electrical force cannot accelerate an electron that occupies a state in a filled band unless the force is strong enough to pull the electron out of the occupied band of states and put it into some state in a band of unoccupied states of high energy. But electrical forces of the usual strength are not nearly strong enough to do that. This is the way wave mechanics explains the distinction between electrical conductors and insulators. In conductors some of the states in a band of permitted energies are not occupied by electrons; in insulators every band is either wholly occupied or wholly unoccupied by electrons.

Before applying these ideas to some examples of real solids, it may be well to rehearse them with the aid of Fig. 11. The electronic

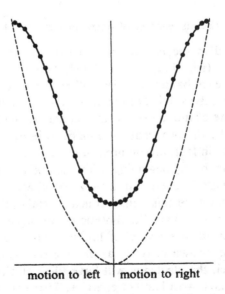

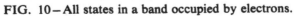
motion to left   motion to right

FIG. 10 — All states in a band occupied by electrons.

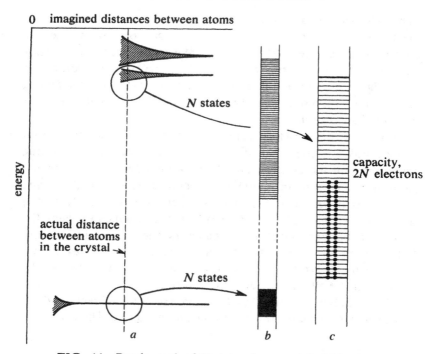

FIG. 11—Bands, each of N states, in a crystal of N atoms.

states permitted in the isolated atoms will be split into bands of permitted states (Fig. 11*a*) as the atoms come together. The number of states in each band (Fig. 11*b*) will be the same as the number of atoms forming the solid. The electrons, all of which would tend to occupy the state of the lowest permitted energy, will be prevented from doing so by the properties summarized in the exclusion principle. They will, in fact, do the next best thing (Fig. 11*c*): they will fill the states in pairs, beginning with that of lowest energy and proceeding upward in energy until they are all accommodated.

A simple example of this behavior is a crystal of lithium. In your imagination start with bare lithium nuclei, arranged in the way the atoms are arranged in a crystal of lithium, and throw in electrons. When you have thrown in twice as many electrons as the number of lithium nuclei, the electrons will have filled all the states corresponding with the lowest band of energies. These are the states that arise from the state that forms the innermost shell on each lithium atom when it is isolated. That energy band is very narrow, and the next band is quite far above it, separated by a wide *energy gap*. The states in this second band are those which arise from the state in the second shell on an isolated lithium atom that the third electron adopts in such an atom.

You now have left as many electrons as nuclei, and when you throw them in, they are forced into states corresponding with the second energy band. Since there are as many of these states as there are nuclei, you could again put in twice as many electrons as nuclei. But since you have only half of that number of electrons left, you will fill only half of the states, each by putting two electrons into it. The occupied states will correspond with the lowest energies in the band, and you will end with a *half-filled energy band* (Fig. 11c).

Our study of the conduction process earlier in this chapter now shows that the electrons in the lower band cannot provide electrical conduction, because that band is completely filled with electrons. But the electrons in the higher half-filled band can all contribute to conduction. In fact, a crystal of lithium is a good electronic conductor, behaving as if it had one free electron per atom. The electron pairs that fill the lower band form the electronic cores of the lithium ions that build the framework of the crystal.

Reasoning of this kind can be used to resolve the second inconsistency described at the beginning of Chapter X—the fact that a sodium chloride crystal, which seems to have an electron-deficient bonding that would make it a metal, is, in fact, an insulator. In sodium chloride the electrons exactly fill each energy band that is occupied at all. The lowest unoccupied energy band is separated from the highest filled band by an energy gap that is sufficiently large to prevent any ordinary electrical force from making the electrons drift.

Beryllium has two electrons in its second shell—one more than lithium. We might predict that the second energy band of a beryllium crystal would be completely filled, providing no more conduction than the first band. But, in fact, beryllium is a metallic conductor, not an insulator.

This seeming discrepancy between theory and experiment can be quickly removed. In beryllium, the third band, next above the second band in energy, is not far enough above that second band to leave an energy gap. In fact, the second and third bands overlap; the bottom of the third band is lower in energy than the top of the second band. The third band, like the second, arises from an energy level in the isolated atom that belongs to a state in the second shell of the beryllium atom's states. States in the same shell often have energies more nearly alike than states in different shells (Fig. 8 of Chapter XI); for this reason the bands derived from states in the same shell will often overlap.

When two bands overlap, some of the electrons that would otherwise seek states in the upper part of the lower band will find states of lower energy in the lower part of the upper band (Fig. 12). Then both bands are incompletely filled, and the material is a metallic conductor.

Another type of material has attracted great scientific and technological interest in recent years. In these materials a filled band and an empty band are separated by an energy gap that is very small. When the gap is small enough, heat energy is sufficient to excite a few electrons from the top of the filled band across the gap and into the bottom of the empty band. Then the "filled" band is

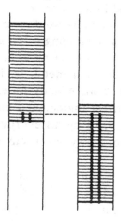

FIG. 12—Overlapping energy ranges of two bands, with both bands incompletely occupied.

not quite filled, and the "empty" band is not quite empty; electrons in both bands can carry a current. Since there are relatively few electrons free to do this, these materials do not carry current as well as the metals and earn the name *semiconductors*. The next chapter will outline some theories that have been developed to explain their curious and useful behavior.

XIV. SEMICONDUCTORS

*The technology which has transformed practical existence is
largely an application of what was discovered by these
allegedly irresponsible philosophers.*

SIR CYRIL HINSHELWOOD,
Address on the Tercentenary of the Royal Society

THE MOST widely heralded studies of solids in recent times are the
investigations and applications of semiconductors. Theory has
suggested experiment and experiment has confirmed theory so
rapidly and so successfully that a new technology has emerged
within the past two decades. Tiny *transistors* now amplify electrical messages where bulky and power-hungry vacuum tubes were
used in the past. Electric power, derived hitherto from the sun's
power through a devious chain of plant growth, combustion, and
steam turbines, can now be generated directly from the sun's light
by *solar cells*. The last few chapters have described much of the
groundwork necessary for understanding these devices.

Semiconductors can be distinguished from metals and insulators
in two ways. In the first place, as their name suggests, their conductivities at ordinary temperature fall between those of metals
and those of insulators. In the second place, their conductivities
increase with increasing temperature, as do the conductivities of
insulators, discussed in Chapter IX. As in metals, however, the
current conducted by semiconductors is not accompanied by a
transport of atomic matter—only by a drift of electrons.

A good way to start picturing the behavior of semiconductors
is to rehearse the *band structures* described in the last chapter.
Figure 1 summarizes the differences that distinguish conductors,

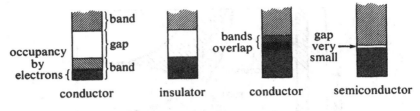

FIG. 1—Four types of band structures.

semiconductors, and insulators. The energy of heat at ordinary temperature is sufficient to excite a few electrons to cross the small energy gap between the filled and unfilled bands in a semiconductor, providing a small electronic conductivity. An increase of temperature rapidly increases the number of electrons that are excited across the gap. This increase in the number of current carriers more than offsets the increase in their scattering described in the last chapter, and therefore the resistance of semiconductors decreases with increasing temperature.

How a small energy gap may arise can be seen especially well in two of the most important semiconducting materials—silicon and germanium. Notice that they fall in the same group of the Periodic Table as carbon. In fact, their crystals have the same arrangement of atoms as the carbon atoms in diamond (Fig. 9, Chapter IV). The arithmetic of electron-deficient bonding described in Chapter VII would therefore give them *saturated bonding* (Fig. 3, Chapter VII) and would suggest that they should be insulators like diamond. But a detailed examination of their energy bands shows that the energy gap between their filled and empty bands is much smaller than the corresponding energy gap for diamond—sufficiently small to make them semiconductors.

Figure 2 describes why diamond differs from silicon and germanium. The energy bands appropriate to the crystal structure of diamond have complicated shapes, but for many purposes they can be summarized as shown in Fig. 2. In diamond, silicon, and germanium, the lower band is completely filled with electrons, and the upper band is empty. At the spacing of the carbon atoms in diamond, the filled and empty bands are separated by a wide gap in energy. At the spacing of silicon and of germanium atoms, the gap becomes very narrow.

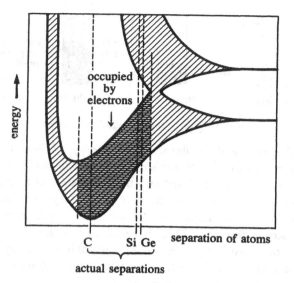

occupied
by
electrons
↓

energy →

C   Si Ge   separation of atoms

actual separations

FIG. 2—The energy gap between the occupied and unoccupied bands in the diamond structure.

Electrons and Holes

Another way of looking at semiconductors like silicon and germanium helps to make clear the relation between the picture of electronic bands and the picture of covalent bonds between neighboring atoms. Almost anyone, thinking about a single covalent bond, thinks of a pair of electrons as staying permanently in the bond between two chosen atoms. But nobody can know that they do stay there. Some pair of electrons is there almost all the time but not necessarily a particular pair. One electron might exchange places with some other in a bond between another pair of atoms in the crystal. When one electron moves one way in the exchange, the other would move in the reverse direction, and there would be no electric current. Since all electrons are alike, there would be no way to tell that the exchange had occurred.

Saying that the lower band in Fig. 2 is filled is the same as saying that all the electrons are in one or another of the electron-pair bonds. For that reason the lower band is often called the *valence band*. The wave for the state of each electron extends throughout the crystal, like the waves of Fig. 18 at the end of Chapter XII, because there is no way to know where any particular electron is, except that it is somewhere in the crystal. Squaring that wave gives the relative probability that you will find the electron in one place or another.

Now suppose that the heat energy in the crystal agitates the atoms sufficiently to knock an electron out of an electron-pair bond. That electron is then free to move around in the crystal. Urged by an electric field, it will move until it drops back into another bond that has also lost an electron. While it is free to move in this way, its state corresponds to an energy in the upper band of Fig. 2, which is therefore often called the *conduction band*. The energy gap between the valence band and the conduction band is the energy required to knock the electron out of the electron-pair bond—the additional energy that the electron must acquire in order to escape and become a *conduction electron*.

It is helpful to describe these activities in terms of particles rather than waves. Figure 3 shows the crystal structure of an

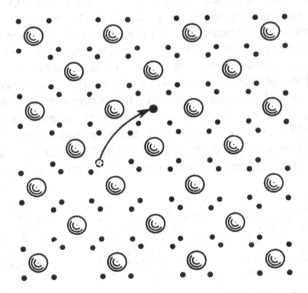

FIG. 3—Thermal excitation of an electron from the valence band to the conduction band in germanium.

imaginary two-dimensional semiconductor that has one feature in common with the structure of silicon and germanium; each atom forms electron-pair bonds with four others. One electron has been knocked out of one of these bonds and is relatively free to roam in the crystal.

In this way the valence band has lost one of its electrons. Since the valence band is then only partly filled, the electrons in states within that band can also conduct electricity. Figure 4 shows how they might do it. An electron in some bond near the bond that has been bereft can move in to reestablish the electron pair. By this act it bereaves its original bond, and a third electron, in still another bond, can move. Thus, conduction takes place by a motion of electrons which successively fill each others' places.

This process is somewhat reminiscent of the motion of protons in a crystal of ammonium dihydrogen phosphate that has a proton deficiency, described in Chapter IX. Here again it is simpler to think of the conduction as taking place by the motion of a deficiency in the opposite direction to that of the particles which fill it. The electron deficiency in the bond has left the region around it with a net positive charge (Fig. 5). Thus, the motion of the electron deficiency corresponds with the motion of a positive charge in a direction opposite to the motion of the electrons, much as the motion of a proton deficiency (Fig. 11 of Chapter IX) was described as the motion of a negative charge (Fig. 12 of Chapter IX).

A mobile electron deficiency like this is often called a *hole in the valence band*, or simply a *hole*. It behaves in many ways like a positively charged particle, which can be pulled about by an electric field. Hence, for each electron excited from the valence band up to the conduction band, two current-carrying particles are

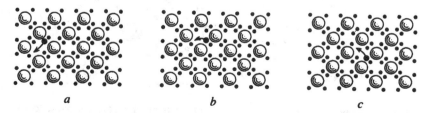

a   b   c

FIG. 4—Conduction in the valence band.

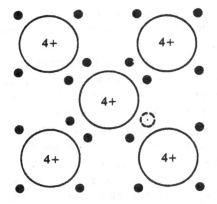

FIG. 5—A net positive charge in the neighborhood of a bond that has lost one electron.

produced: a conduction electron and a hole. When an electric field is applied to the semiconductor, these two types of particles move in opposite directions because they have opposite charges. But, again because they have opposite charges, those motions contribute electric currents which are in the same direction and add to each other.

Impurity Semiconductors

So far, the discussion has dealt with conduction electrons and holes that are freed by the heat energy, or thermal vibrations, of the atoms in the semiconducting crystal. In ammonium dihydrogen phosphate, mobile proton deficiencies can arise also from impurities deliberately introduced into the crystal, as Chapter IX described. Silicon and germanium likewise can play host to impurities that will introduce extra current-carrying particles. By choosing those impurities properly and adding them to a pure crystal deliberately, we can introduce either extra conduction electrons or extra holes.

A glance at the Periodic Table suggests which impurities to pick. Germanium and silicon fall in the column of the table designated Group IV; each element in that column has four *valence electrons* in its outermost occupied shell. The elements in Group III, immediately to the left in the table, have only three such electrons, and a crystal of silicon or germanium will accept a limited number of atoms of those species in its crystalline sites, replacing atoms of silicon or germanium.

By offering one fewer electron to complete the bonding of the structure, each boron or aluminum atom introduces an electron deficiency — a hole in the valence band. Conversely, an atom of a species in Group V — phosphorus or arsenic, for example — provides one more electron than the valence bonds can accept, and that electron becomes a conduction electron. The behavior of these two sorts of impurities is schematized in Fig. 6. Impurities like arsenic are called *donors* because each atom donates one conduction electron; impurities like aluminum are called *acceptors* because each atom contributes an electron deficiency that will accept one electron.

No matter how the conduction electrons and holes are introduced — whether by heat or by impurities — they are quite mobile in the crystal, nearly as mobile as the free electrons in a metal. But since they are very much less numerous than the electrons in a metal, a semiconductor always conducts electricity much less readily than a metal. Even though the conductivity of a semiconductor increases with its temperature, while that of a metal decreases, the conductivity of the semiconductor never exceeds that of the metal.

Visualize the mobile conduction electrons and holes as diffusing fairly freely through the semiconductor, much as molecules of perfume diffuse through the air. When a conduction electron meets a hole, the electron may enter the hole; in other words, the electron may complete the incomplete electron-pair bond. But for every

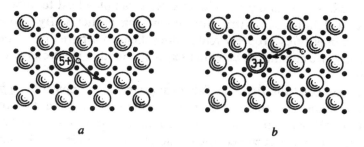

a   *b*

FIG. 6 — (*a*) A donor impurity. (*b*) An acceptor impurity.

such pair of current-carriers that combine with each other, the ever-present thermal agitation produces another pair. In short, a pure crystal contains an *equilibrium number* of hole-electron pairs.

When an impurity—arsenic, for example—is added to the crystal, it no longer has the same number of conduction electrons and holes, because arsenic introduces only extra conduction electrons. The increased concentration of conduction electrons makes it more likely that some one of them will combine with a hole. Thus, any impurity that introduces extra conduction electrons into a semiconductor automatically reduces the number of holes in it, and vice versa. Nevertheless, the impurity will still increase the total number of current carriers and thus increase the conductivity of the material.

When both kinds of impurities are put into the crystal, their effects tend to cancel; they will cancel exactly if there are exactly the same number of atoms of each impurity. The two types of impurities have subtractive, not additive, effects, and the impure material can behave much like the pure material. The impurities may even decrease the conductivity a little by increasing the scattering of the current-carriers in the way described in the last chapter.

The *p-n* Junction

In the principal technological uses of semiconductors, the two types of impurities are introduced into adjacent parts of a single crystal of the material. For example, the crystal of germanium whose arrangement of atoms is sketched in Fig. 7 contains on its left side a small amount of an acceptor impurity and on its right side a small amount of a donor impurity. Thus, the left side of the crystal is made rich in mobile holes and the right side in mobile electrons.

A semiconductor in which mobile electrons outnumber holes is said to be of *n* type because its majority carriers bear a *negative* charge. A *p*-type semiconductor is named for the *positive* charge of the holes that constitute its majority carriers. When a *p*-type and an *n*-type semiconductor are joined (Fig. 7), the region in which the type changes is called a *p-n junction*.

Initially you might expect to find the mobile carriers only in the

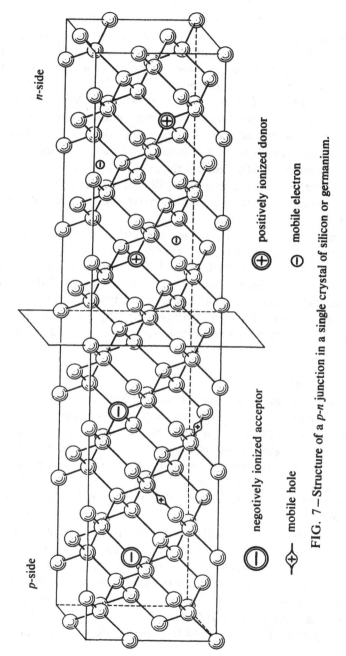

FIG. 7 — Structure of a *p-n* junction in a single crystal of silicon or germanium.

⊕ positively ionized donor

⊖ mobile electron

⊖ negatively ionized acceptor

⊕ mobile hole

n-side

p-side

regions occupied by their parent ions (Fig. 8). In order to complete its bonding function in the structure, each acceptor atom on the left will acquire an electron from some neighboring bond in the structure, becoming a negative ion and leaving a positive hole to wander. On the right each donor atom will become a positive ion, freeing its extra electron to wander likewise.

But the arrangement cannot look like that for long. Some holes will wander across the junction into the region on the right, where they will meet an excess of electrons to combine with them. Similarly some electrons will wander into the region on the left, where they will be captured by holes and become bound. This wandering is a diffusion of holes and electrons in opposite directions, similar to the diffusion of atoms described in Chapter IX. As Fig. 6 in that chapter pictured, mobile particles tend to diffuse from regions where their concentration is high into regions where their species is less numerous. In this instance, the diffusion of holes and electrons could be visualized as in Fig. 9.

In the more familiar cases of diffusion, the process continues until the concentrations of the diffusing species have become uniform throughout the accessible space. In this case, however, there is a special reason that the holes and electrons cannot equalize their concentrations on both sides of the junction. Each impurity atom made its contribution to those charge-carriers by gaining or losing an electron and becoming an ion bearing a net charge that is frozen in place in the crystal. When the mobile charges diffuse away, they leave those immobile charges behind them. The accumulation of ionic charge, of opposite sign to the mobile charge, tends to pull the mobile charge back. Hence, near the junction, an equilibrium is accomplished. After diffusion has progressed enough

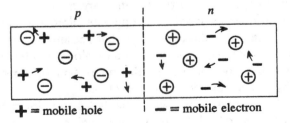

$+$ = mobile hole $\qquad$ $-$ = mobile electron

FIG. 8 — Ionized impurities near a *p-n* junction.

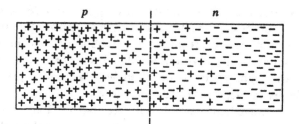

FIG. 9 — Diffusion across a *p-n* junction.

to leave a fixed ionic charge, that charge not only opposes further diffusion but also attracts a compensatory flow in the other direction.

Figure 10 summarizes the condition of the junction when these processes have reached equilibrium. A *double layer* of electrical charge is constructed near the junction because diffusion of mobile charges has left the fixed ionic charges partly uncompensated, and the double layer opposes further diffusion. The opposition is equivalent to a hill that the mobile charges must climb in order to leave the region from which they diffuse — a hill that faces in opposite directions for holes and for electrons.

The Junction Rectifier

Picture now what happens when that equilibrium is disturbed by attaching the opposite poles of an electric battery to the opposite ends of the junction-bearing crystal. Urged by the electromotive force of the battery, holes and electrons acquire drift velocities that add to the velocities of their random motions, as the last chapter described. Since the holes bear a positive charge and the electrons a negative charge, their drift velocities are in opposite directions. But opposite charges moving in opposite directions give electric currents in the same direction. For that reason we can focus our attention on what happens to the holes alone, confident that the electric current contributed by the holes will be approximately doubled by the current due to the electrons.

Figure 11*a* shows again what the holes are doing before the battery is connected. Holes from the *p* region are diffusing to the right across the junction, despite opposition by the charge on the

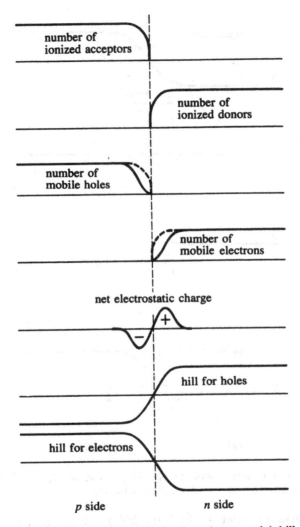

number of
ionized acceptors

number of
ionized donors

number of
mobile holes

number of
mobile electrons

net electrostatic charge

+

−

hill for holes

hill for electrons

p side   *n* side

FIG. 10—Charge distribution and electrostatic potential hills at a *p-n* junction.

positive ions, to give an electric current I_f. Holes thermally generated in the *n* region on the right are contributing an exactly compensating current I_g. The holes giving I_f are climbing up the hill. The holes that give I_g wander randomly in the *n* region until by chance they come to the brow of the hill and slide down.

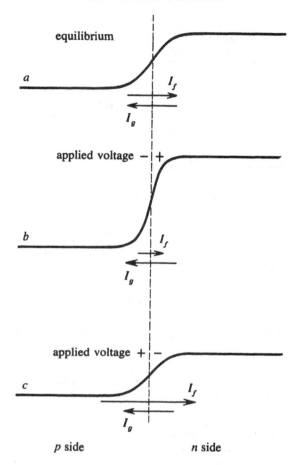

FIG. 11 — Potential hills and currents of holes across a *p-n* junction (*a*) at equilibrium and (*b* and *c*) with applied voltages.

Of course, in order to climb the hill and contribute to I_f, a hole must have enough energy to reach the top. Any less energetic holes attempting the ascent will slide back. The size of I_f will therefore depend on the height of the hill and on how many holes have enough energy to surmount it. The number of holes that can reach the top of the hill can be found by the methods of statistical mechanics, the analytical tool that yielded the laws of the behavior of perfect gases mentioned in Chapter II. Figure 12 shows the general form of the result. The fraction of holes with enough energy to climb the hill increases rapidly as the height of the hill decreases.

An electric battery connected to the crystal changes the height of the hill. If the positive terminal of the battery is connected to the n side of the junction and the negative terminal to the p side, the battery assists the positive ions on the n side to repel the holes. In other words, it increases the height of the hill (Fig. 11b). If the battery connections are reversed, the hill is lowered (Fig. 11c).

Changing the height of the hill makes little change in I_g. That current depends on the rate at which holes are thermally generated in the n region and survive their hostile environment long enough to diffuse to the brow of the hill. A hill of any height will urge ahead all holes that stray to its brow. But changing the height of the hill can make a large change in I_f. A lower hill allows many more holes to climb it, and a higher hill produces less change (Fig. 12).

The resulting net current is plotted in Fig. 13. It increases rapidly with the voltage of the battery when the battery is connected so as to reduce the height of the hill. It increases less rapidly in the other direction and finally becomes unchanging when the battery raises the height of the hill. That final, unchanging *saturation current,* produced when the hill is so high that almost no hole has enough energy to climb it, is just the current I_g.

Notice that the piece of germanium containing the p-n junction is

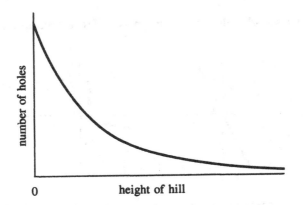

FIG. 12 – Increase in the number of holes able to climb the hill of Fig. 11 as the hill is lowered.

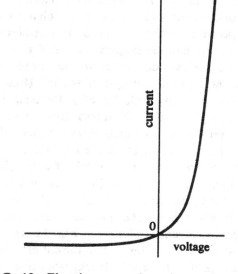

FIG. 13 — Electric current through a *p-n* junction.

far from being obedient to Ohm's law, mentioned in the last chapter. The rule that the current is proportional to the voltage, so useful in describing electrical conduction by metals, is inapplicable here. Figure 14*b* makes clear why the *p-n* junction is called *non-ohmic*.

Notice also why the *p-n* junction is called a *rectifier*, the name

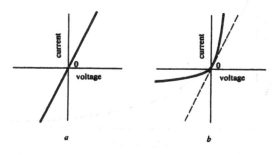

FIG. 14 — (*a*) Ohm's law. (*b*) A *p-n* junction.

given to any device that converts an alternating current into a direct current. When the junction is connected to a source of alternating voltage, it behaves as it would if it were connected to a battery whose connections were rapidly reversed. It passes a large current during half of each alternation and a very small current during the other half (Fig. 15). The current is like a direct current with a ripple superimposed on it. For many purposes—charging storage batteries, for example—the ripple is unimportant. When it is undesirable, the ripple can be ironed out by other electric devices.

The Solar Cell

When the *p-n* junction is used as a rectifier, the equilibrium of forward and backward currents described in Fig. 11 is upset by a

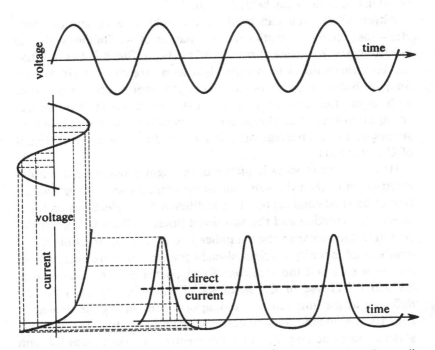

FIG. 15—An alternating voltage, producing through a *p-n* junction a direct current with a superimposed ripple.

voltage applied across the junction. There are other ways to upset that equilibrium; one of the most useful is to illuminate one side of the junction with light. As the light is absorbed, it ejects electrons from the bonds in the semiconductor and thus adds more hole-electron pairs to those produced by the thermal agitation.

There are two viewpoints from which you can see how light produces the additional hole-electron pairs. Thinking of light as an electromagnetic wave, you can imagine that the oscillating electrical force of that wave pushes the bonding electrons rapidly back and forth because they bear an electrical charge. The resulting picture will be somewhat like that of the infrared light pushing the chloride ions in a sodium chloride crystal (Fig. 2 of Chapter IX). Here, however, the light is oscillating a hundred times faster and shakes the electrons hard enough to tear them loose from their positions, leaving holes behind them.

Alternatively, you can think of the light as a stream of particles — the photons mentioned in Chapter X — colliding with the electrons and knocking them out of position. But it is best to fuse the two viewpoints — to picture light as a stream of wave bursts. Such a burst — a photon — can exchange energy and momentum with an electronic wave burst when the wavelengths of the bursts are appropriate. Then the picture is reminiscent of the collision of an electron with an irregularity in a crystalline arrangement (Fig. 9 of Chapter XIII).

However the process is pictured, the light produces more hole-electron pairs than the semiconductor contains in its usual condition of thermal equilibrium. The additional freed electrons on the n side of the junction and the additional holes on the p side are hardly noticeable, because their number is only a small fraction of the numbers of majority carriers already present. But the holes added on the n side and the electrons freed on the p side make a large fractional increase in the numbers of the minority carriers. By diffusing to the junction and sliding down the hills diagrammed in Fig. 10, the added minority carriers increase I_g (Fig. 11), while I_f stays almost unchanged, and a net electric current passes through the cell.

Thus, the operation of the solar cell depends on the capture, by the junction, of freed electrons drifting from its p side. But on that side they are greatly outnumbered by holes. The odds are heavy that an electron diffusing toward the brow of the hill will combine with a hole before it can escape through the junction. For this reason, the only useful electrons are those that are freed close enough to the junction to reach it before they are lost by combining with holes.

In other words, the light that energizes the solar cell will be wasted if it is absorbed farther from the junction than a *diffusion length*—the average distance that an electron diffuses before a hole captures it. Solar cells are therefore designed as shown in Fig. 16. The *p* region is given a large area in order to accept as much light as possible and a thickness of about 10^{-4} inch in order to use the accepted light efficiently.

The *n-p-n* Transistor

You have now met all the kinds of behavior necessary for understanding the simplest type of *transistor*—the name given to several sorts of solid devices that can accomplish most of the electronic duties long performed only by vacuum tubes. To make

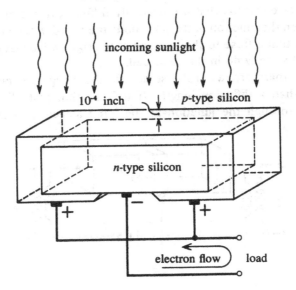

incoming sunlight

10^{-4} inch   *p*-type silicon

n-type silicon

+   −   +

electron flow   load

FIG. 16—The solar cell.

an *n-p-n* transistor, two *p-n* junctions are introduced back to back into a single crystal of germanium or silicon. The three distinguishable regions so produced are connected to sources of constant voltage, to the source of signals to be amplified, and to the electric circuits that will accept the amplified signals (Fig. 17).

A battery of low voltage in the input circuit is connected across the *n-p* junction on the left, with its negative pole to the *n* side and its positive pole to the *p* side. Subjected to voltage of that polarity, the junction passes the relatively large current shown at the right side of Fig. 13. Furthermore, the varying signal voltage, superimposed on the constant voltage of the battery, produces a large variation in the current through the junction. Thus, the *p* region in the middle of the transistor receives from the left a large input of conduction electrons, varying from instant to instant with the signal voltage.

The battery of high voltage in the output circuit is connected across the *p-n* junction on the right with its negative pole to the *p* side and its positive pole to the *n* side. That junction is therefore normally passing the very small current shown at the left side of Fig. 13. But that current is small only because it is limited by the number of electrons that are normally diffusing to the brow of the hill. When the junction at the left floods the *p* region with electrons, all those that diffuse to the junction at the right will coast down hill also, just as they do in the solar cell.

Again, therefore, as in the solar cell, the *p* region should be no thicker than a diffusion length. If it is so thin that all electrons injected through the junction at the left succeed in reaching the

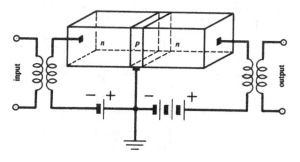

FIG. 17—The *n-p-n* transistor.

junction at the right, the signal current through the output circuit will be as large as the signal current through the input circuit.

What has been gained by going to all this trouble to produce an output current that can do no better than equal the input current? The answer comes from examining the voltage along with the current. When a small fluctuation of the signal voltage across the left junction produces a current fluctuation, then the same current fluctuation, passing through the right junction, must be accompanied by a large voltage fluctuation across it (Fig. 13). In other words, the voltage has been amplified without losing much current. Since the power dissipated in a circuit is proportional to the current through it times the voltage across it, the transistor amplifies the power available to the output circuit.

Power cannot be had for nothing; where does it come from? It comes from the battery across the junction at the right. The transistor converts power from that battery into power in the output signal by triggering the battery with the little input signal. In a carefully made transistor the output signal can be given as much as a 10^5 times the power of the input signal.

XV. MAGNETS

. . . I've seen
Those Samothracian iron rings leap up,
And iron filings in the brazen bowls
Seethe furiously, when underneath was set
The magnet stone.

LUCRETIUS, *De Rerum Natura**

HOW DOES matter act where it is not? Throughout recorded history, natural philosophers have been puzzled by that question. Divine ordinances were invoked to keep the planets in their places until Newton invoked the even grander idea of universal gravitation acting between the planets and the sun without contact. But the actions accomplished by magnets must have seemed too local and selective to engage the attention of the gods.

Only a particular mineral, *lodestone,* appeared to be endowed by Nature with these remarkable abilities. It was able to endow other things in turn but only if they were made of iron. For over 2,000 years, the explanation of this behavior ran the gamut of intellectual ingenuity. Writing about 60 B.C., Lucretius suggested,

> . . . stream there must, from off the lode-stone seeds
> Innumerable, a very tide, which smites
> By blows that air asunder lying betwixt
> The stone and iron.*

After the air has been thus removed,

> . . . the primal germs
> Of iron, headlong slipping, fall conjoined
> Into the vacuum, . . .*

*Of the Nature of Things, a metrical translation by W. E. Leonard (New York, Dutton, 1916; London, Dent, 1916), pp. 290–91. See also Everyman's Library edition, 1921. Used by permission.

William Gilbert of Colchester, whose experimental assiduity 1600 years later laid the foundations of the present understanding of electricity and magnetism, still felt need of an "effluvium" to explain what he observed. And less acute observers continued to propagate legend and nonsense, such as,

The lodestone onely by the affriction of Garlick, amits its verticity, and neglects the pole, conserving to itself, in the meantime, its peculiar forme, materiell constitution, and all other dependent proprieties. The reason, because Garlick is the lodestone's proper Opium, and by it that spirituell sensation in the magnet is consopited and layd asleep.*

The philosophically minded reader of this book will already have noticed that today the problem of action at a distance seems less worrisome than it seemed in the past. The present pictures of matter even suggest that it acts only at a distance. On the atomic scale, the particles of which matter is made — the nuclei and their associated electrons — affect one another without contact by electrostatic forces that act over the short distances within an atom much as those forces act over longer distances between charged pith balls.

More worrisome today is the need, still with us, to think of different kinds of forces. The gravitational force, which depends on the magnitudes of the gravitating masses, seems to be distinct from the electrostatic force, which depends on the magnitudes of the interacting charges. A grand design that will bring these two sorts of forces into a single picture still eludes us.

Elementary Magnets

Magnetic forces, however, can be given a place in the electrical picture. The electromagnetic theory that unifies electrical and magnetic phenomena has developed over the past 200 years through the experiments and interpretations of many people, among whom Hans Oersted, André Ampère, Michael Faraday,

---

*Jean Baptiste Van Helmont, *De magnetica vulnerum curatione* (1621), translated by Walter Charleton as *A Ternary of Paradoxes* (1650) and quoted in Paul F. Mottelay, *Bibliographical History of Electricity and Magnetism* (Philadelphia, Lippincott, 1922). Van Helmont may perhaps be pardoned for this repetition of myth in view of his contemporaries' attribution of many unusual powers to garlic, including the power to ward off vampires.

and James Maxwell were outstanding. Today it is clear that magnetism is always a manifestation of moving electrical charges.

Positive and negative electrical charges seem to be a primary property of the particles of which matter is composed. But apparently magnetic poles, north and south, are not. All the magnetic poles so far investigated have appeared in equal and opposite pairs, and the magnetic dipole formed by an associated pair can always be ascribed to the motions of the particles that bear the primary electrical charges.

This idea was applied in Chapter XI to the spinning electron. If the electron is visualized as an electrically charged ball that is spinning about an axis through it, then the electromagnetic theory predicts that the electron will behave like a tiny magnet (Fig. 10 of Chapter XI). In fact, spectroscopic experiments in which magnetic forces are deliberately applied to matter suggest that each electron must exhibit a magnetic dipole; it is an *elementary magnet*.

Since every piece of matter contains spinning electrons, it contains elementary magnets. But usually the piece taken as a whole does not behave like a magnet, because the elementary magnets in it are turned in different directions and cancel one another's magnetic effects. One reason for that cancellation is that electrons usually occupy their permitted states in pairs; in each pair one electron's spin is up and the other's is down (Fig. 13 of Chapter XI). Such a pair contributes no net magnetic dipole.

In some materials, however, the elementary magnets do not cancel one another's effects; they add to one another's effects. Iron, cobalt, nickel, and the lodestone of the ancients are among such materials that have long been known, and more are constantly being discovered. Because iron provides the most familiar examples, the magnetic property of this class of materials is called *ferromagnetism*.

Domains

But pieces of iron are still not magnets until they have been magnetized. As even the ancients knew, they must be put near something that will exert magnetic forces on them—something that is already magnetized, like the lodestone. To be sure, Gilbert had noticed that a bar of iron can be magnetized by being hammered while held so that its two ends point north and south. But Gilbert had also noticed that the earth itself is a great magnet and had thereby explained the action of a magnetic compass needle. Today we know that a coil of wire carrying electric current (Fig. 10a of Chapter XI) will also serve.

Early in this century Pierre Weiss hit upon a satisfactory explanation of how a ferromagnetic material becomes magnetized in the presence of magnetic forces applied from outside. He suggested that each piece of the material consists of regions already magnetized and that the direction of their magnetization differs from one region to another so as to cancel out in the whole piece. When the piece is subjected to magnetic forces from outside, the direction of magnetization within each region turns toward alignment with those forces, and a net magnetization appears in the material.

Later work confirmed Weiss' idea that a piece of ferromagnetic material is divided into domains of spontaneous magnetization so disposed as to make the piece appear unmagnetized. Figure 1 shows a somewhat idealized arrangement of domains of that kind, appropriate to a single crystal of a ferromagnet. The material lowers its energy by changing from a grossly magnetized condition (Fig. 1a) to a condition in which many magnetized domains cancel one

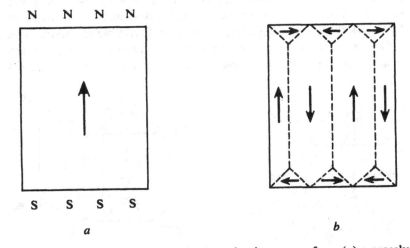

FIG. 1 – A ferromagnetic material lowering its energy from (a) a grossly magnetized condition by (b) forming domains.

another (Fig. 1*b*), and the material does not appear to be magnetized. Arrangements of domains similar to that in Fig. 1*b* have actually been seen under a microscope in an iron crystal, by observation of their effects on magnetic powders suspended in a liquid and smeared on a polished surface of the crystal.

It turns out that a domain cannot choose the direction of its magnetization at random; it must choose that direction from among a very few easy directions of magnetization in the crystalline structure of the material. For example, Fig. 2*a* shows the six easy directions in iron and Fig. 2*b* shows the eight directions in nickel in relation to a unit cell of their crystal structures. In a domain made of many such unit cells, the magnetization appears along only one or another of these directions, and the direction of magnetization can be made to depart from these few only by very large magnetic forces.

For some time after Weiss made his suggestion, the process of magnetizing a piece of iron was visualized as a sudden "flopping" of the direction of magnetization within the domains, from one of the easy directions to another one more favorably disposed to the magnetizing force. More recently, it has become clear that the magnetization within a domain does not "flop" in so abrupt a fashion. Instead, the magnetizing force makes the favorably disposed domains grow at the expense of those less favorably disposed (Fig. 3). The wall between two adjacent domains moves more or less smoothly, providing a larger region of favorable magnetization.

On an atomic scale the wall itself is not abrupt; the elementary magnets in the wall also change their directions smoothly. Figure 4

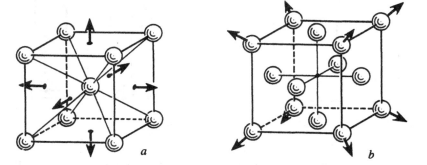

FIG. 2—Easy direction of magnetization in (*a*) iron and (*b*) nickel.

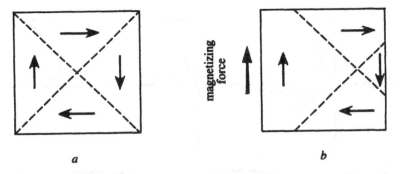

FIG. 3 — Magnetization of (*a*) an unmagnetized ferromagnet by (*b*) moving domain walls.

suggests how the directions of the elementary magnets are probably disposed in a domain wall. The wall moves by a smooth and orderly change in the directions of the elementary magnets. Urged by a magnetizing force applied from outside, the direction of magnetization of the elementary magnets rotates about a line perpendicular to the wall, and thus the wall moves.

The Critical Temperature

With the aid of this picture of moving domain walls, the process of magnetizing a ferromagnet can be traced in the way shown in Fig. 5. As the magnetizing force increases, the magnetization

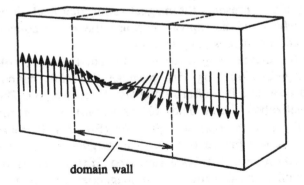

domain wall

FIG. 4 — Smooth change in direction of magnetization within a domain wall.

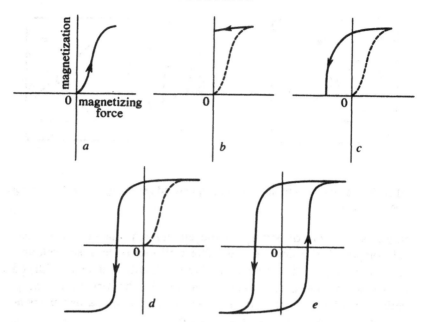

FIG. 5 — The magnetization of a ferromagnet with changing magnetizing force.

increases more and more strongly (Fig. 5a) because more domain walls are enabled to move. But the magnetization cannot increase indefinitely; when there are no walls left to be moved, the force can do little to increase the magnetization further. That final *saturation magnetization* represents the magnetization that each domain had already, which did not appear because the many domains were magnetized in compensating directions.

When the magnetizing force is removed, the piece still shows some magnetization (Fig. 5b) because the walls cannot easily return to positions that afford complete compensation. The apparent magnetization can be made to vanish only by exerting a magnetizing force in the opposite direction (Fig. 5c). Increasing that reverse force still further (Fig. 5d) can magnetize the piece in the other direction until it is again saturated. If this process is continued by cycling the direction of the magnetizing force, the plot of magnetization against force (Fig. 5e) traces a loop called a *hysteresis loop*.

In the technological uses of magnets, the form of that loop is extremely important. The magnetization that remains when the magnetizing force is removed, for example, is a measure of how strong a permanent magnet can be made of the material. The area of the loop is a measure of how much energy will be lost as heat if the material is used to make cores in transformers. Much effort is steadily perfecting the manufacture of diverse magnetic materials especially suited to their diverse uses. The schedule of how the substance is heated and cooled, the regimen of mechanical treatment and mistreatment – all such fabricating arts affect the product for good or ill.

Of more fundamental concern, however, is the saturation magnetization of the material. Since that represents the spontaneous magnetization within each domain, it is a property of the substance that is independent of how it is fabricated into pieces. Again, as with electrical conduction, the clearest insights come from examining how the property varies with the temperature.

The form of that variation in nickel is shown in Fig. 6. As the metal is heated, the saturation magnetization declines, and at a *critical temperature* (631 degrees absolute, or 358 degrees centigrade), it falls precipitously. Above that temperature, the substance is no longer a ferromagnet; it behaves like many other metals.

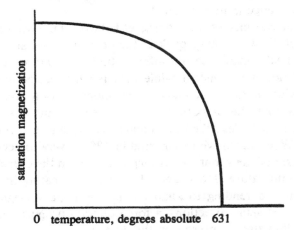

FIG. 6 – Change of saturation magnetization with temperature in nickel.

When it is cooled again, the magnetization within the domains reappears and traverses the same curve in the opposite direction. In other words, the spontaneous magnetization of the material varies simply and reversibly with the temperature and vanishes above a critical temperature. Iron behaves similarly but with a different critical temperature (1043 degrees absolute, or 770 degrees centigrade).

The Internal Field

This behavior is somewhat reminiscent of melting. When a solid material melts, the orderliness characteristic of crystallinity abruptly disappears and the crystals fall apart into a disorderly liquid. The vigor of the atomic vibrations described in Chapter II becomes sufficient to overcome the forces tending to hold the atoms in the orderly arrays described in Chapter IV.

The analogy between the critical temperature at which a crystal melts and the critical temperature at which ferromagnetism disappears suggests examining the heat capacity of a ferromagnet. Heating an ordinary solid will raise its temperature progressively up to its melting point. Then, as the solid melts, enough heat must be supplied to melt it completely before the temperature of the liquid will rise further. In other words, at that melting temperature the heat capacity of the material is infinitely large; supplying heat makes no change in its temperature.

The heat capacity of a ferromagnet behaves somewhat similarly (Fig. 7); it shows a sharp peak at the critical temperature. To be sure, the heat capacity is not infinite; the temperature of the material will not stay constant while heat is supplied. Nevertheless, the behavior of the heat capacity is anomalous—a conspicuous departure from the smooth behavior observed in other metals, which the models described in Chapter III explain so successfully.

It was Weiss again who suggested in 1907 a way to account for the variation of the spontaneous magnetization in the domains with varying temperature. He imagined a force originating in the material itself and tending to align all the elementary magnets in a domain in the same direction—a force directly proportional to the magnetization already present in the domain.

Notice how neatly that idea can explain the behavior di-

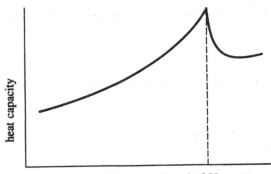

FIG. 7 – Change of heat capacity with temperature in nickel.

agrammed in Figs. 6 and 7. The heat vibrations tend to disturb the alignment of the elementary magnets, increasingly as the temperature rises. The more that alignment is disturbed, the lower is the spontaneous magnetization. But, according to Weiss, the force tending to align the elementary magnets is itself proportional to the extent of alignment that it achieves. Hence, as the temperature increases, the magnetization decreases calamitously; as the temperature decreases, the magnetization again lifts itself "by its own bootstraps." At the critical temperature, the heat capacity displays a peak because a small increase of temperature induces a large increase of disorder in the arrangement of the elementary magnets. For disordering that arrangement, heat is required, just as heat is required to disorder the atomic arrangement in a melting solid.

In order to explain the origin of the internal force that Weiss imagined, it seemed natural at first to turn to the magnetic forces between the elementary magnets. The discussion in Chapter VI of the van der Waals forces between atoms has pictured ways (Fig. 3 of Chapter VI) in which those magnets might tend to align one another. If the magnets were arranged in strings, for example, they would adopt a head-to-tail arrangement (Fig. 8a) of the sort required rather than an arrangement (Fig. 8b) that has no net magnetization.

Unfortunately, when numbers were put into Weiss' theory, it turned out that magnetic forces between the elementary magnets are 10^3 times too small to account for the measured behavior of

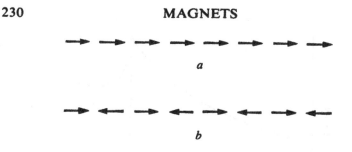

FIG. 8—The arrangement of magnets a has a lower energy than the arrangement b.

ferromagnets. The *Weiss internal field*, so necessary to the picture of ferromagnetic behavior, had to be accepted for many years without explanation of its origin.

It was clear that the needed factor of 10^3 could not be found in the magnetic forces between the elementary magnets. And it was also clear that electrostatic forces operating on the electronic ingredients might have the needed magnitude. Is there a way for electrostatic forces to align the spinning electrons? Werner Heisenberg was the first to recognize that wave mechanics offers a positive answer to this question.

In order to describe his answer, we would need to travel more deeply than this book has traveled into wave mechanics. Here we have spoken of waves for each electron separately, as if we could lay hands on it and label it. But when many electrons are crowded into a solid, there is no way to distinguish one from another. In such a phenomenon as ferromagnetism, the indistinguishable electrons are interacting so strongly that they cannot be studied as if they were separable.

The principles of wave mechanics have been developed to treat the entire assembly of electrons as a single physical system. But as so often happens in physical science, the application of those principles to special cases is difficult. Much of the present research in the physical theory of solids is devoted to perfecting methods for applying wave-mechanical principles to elucidate the collective behavior of electrons.

APPENDIX. SCALES OF ENERGY

MANY different units of energy are used in different contexts, and it is helpful to know a few of the relationships between them.

The classic unit of energy used by physicists is the *erg*, defined as one *dyne-centimeter*. Thus, one erg is the energy expended when one dyne of force moves something through one centimeter of distance. One *dyne* is the force that gives to one gram of mass an acceleration of one centimeter per second per second—in other words, the force that will increase the velocity of the mass by one centimeter per second within one second. The force exerted by the earth on a mass of one gram is 980 dynes, and that is, therefore, the weight of one gram of mass.

The classic unit of heat is one *calorie*, defined as the amount of heat required to raise the temperature of one gram of water through one degree centigrade. James Joule found (Chapter II) that the energy equivalent to one calorie is about 4.2×10^7 ergs. For many purposes, it has been found convenient to define larger units, both of energy and of heat. One *joule*, for example, is defined as 10^7 ergs (whence one calorie equals 4.2 joules), and one *kilocalorie* (a unit widely used by chemists and often abbreviated to kcal) is defined as 10^3 calories.

In problems of atomic physics, it is often convenient (as in Fig. 1 of Chapter XIII) to use for the unit of energy the *electron-volt*, defined as the kinetic energy that is acquired by an electron when it is accelerated through one volt of difference in electrical potential. That unit is equivalent to 1.6×10^{-12} erg—only a tiny fraction. But an atom is very small; a gram-atomic weight of most solids occupies roughly one cubic inch and contains Avogadro's number of atoms (Chapter II), approximately 6×10^{23}. Hence, an energy of nearly 10^{12} ergs (actually about 23 kcal) would be needed to raise the energy of one electron in all those atoms by one electron-volt.

INDEX

Acceptor impurities, 206
Adamantane, 41
Aepinus, Franz, pyroelectric effect, 6
Aliphatic compounds, relation of diamond to, 41
Alkali metals, 78
Alloys, electrical resistance of, 193
Alum: shape of crystal, 50–51; symmetry, 54; Hooke's speculation on structure, 98–99
Aluminum: heat capacity, 24, 25; oxide, 45; ions in beryl, 114; as acceptor impurity in semiconductors, 206
Aluminum potassium sulfate, *see* Alum
Ambivalence of transition elements, 78
Ammonia, bonds in, 87–88
Ammonium ion, 106
Ammonium chloroplatinate, crystal structure, 106–07
Ammonium dihydrogen phosphate, 129 ff.
Ampère, André, electromagnetic investigations, 221
Amplification by *n-p-n* transistor, 217–19
Anharmonic oscillators, 16
Anisotropy, 58–59
Aromatic compounds, relation of graphite, 42
Arrow, symmetry, 60
Arsenic as donor impurity in semiconductors, 206
Atomic weights: related to heat capacities, 13; related to valencies, 76; sequence in Periodic Table, 77
Atomistic explanation, 8 ff.
Atoms: binding between, 70 ff.; planetary theory, 80 ff.; wave-mechanical theory, 151 ff.; ionization energies, 181–82; *see also* Molecules, Electrons, Wave mechanics
Avogadro, Amedeo, number of molecules per gram-molecule, 13
Axes of symmetry, *see* Symmetry

Bands: energy, 177 ff., 187 ff., 200–03; distinguishing insulators, conductors, and semiconductors, 194–203; half-filled, 197; overlapping, 198; separated by small gap, 198, 200 ff.; valence and conduction, 202–03
Barium: in benitoite, 114; as impurity in ammonium dihydrogen phosphate, 132
Barium titanate, 68
Barlow, William, on crystal structure of sodium chloride, 103–04
Benitoite, crystal structure, 114
Benzene, vapor, 42
Beryl, crystal structure, 114
Beryllium ions: in beryl, 114; occupancy of electronic states in metallic, 197–98
Black-body radiation, 21
Bloch, Felix, theory of metals, 174 ff.
Bohr, Niels: model of atoms, 81, 156; quoted, 165
Boltzmann, Ludwig, and statistical mechanics, 14
Bonds, interatomic: Newton's insight into, 1, 70; in diamond, 40, 88, 92, 94, 108, 201; in graphite, 40; ionic, 71 ff., 114; van der Waals, 73 ff., 93, 108, 114; double, 76–77; by electron-sharing, 83 ff., formal rules, 86–88; donor-acceptor, 88, 109; covalent, 88 ff., 170 ff.; one-electron, 90–91; electron-pair, 91; electron-deficient, 91–93, 136; metallic, 91 ff., 135, 170 ff.; saturated, 92; extreme types, 93–94; mixed, 93 ff., 102–03, 111, 136; ion-dipole, 95–97; hydrogen, 97, 116, 128 ff.; tetrahedral, 108 ff.; directional, 108; between organic molecules, 114; in minerals, 114
Born, Max, quoted, 12
Boron: heat capacity, 14, 21, 24, 26; as acceptor impurity in semiconductors, 206

A CATALOG OF SELECTED
DOVER BOOKS
IN ALL FIELDS OF INTEREST

A CATALOG OF SELECTED DOVER
BOOKS IN ALL FIELDS OF INTEREST

100 BEST-LOVED POEMS, Edited by Philip Smith. "The Passionate Shepherd to His Love," "Shall I compare thee to a summer's day?" "Death, be not proud," "The Raven," "The Road Not Taken," plus works by Blake, Wordsworth, Byron, Shelley, Keats, many others. 96pp. 5³⁄₁₆ x 8¼. 0-486-28553-7

100 SMALL HOUSES OF THE THIRTIES, Brown-Blodgett Company. Exterior photographs and floor plans for 100 charming structures. Illustrations of models accompanied by descriptions of interiors, color schemes, closet space, and other amenities. 200 illustrations. 112pp. 8⅜ x 11. 0-486-44131-8

1000 TURN-OF-THE-CENTURY HOUSES: With Illustrations and Floor Plans, Herbert C. Chivers. Reproduced from a rare edition, this showcase of homes ranges from cottages and bungalows to sprawling mansions. Each house is meticulously illustrated and accompanied by complete floor plans. 256pp. 9⅜ x 12¼.
0-486-45596-3

101 GREAT AMERICAN POEMS, Edited by The American Poetry & Literacy Project. Rich treasury of verse from the 19th and 20th centuries includes works by Edgar Allan Poe, Robert Frost, Walt Whitman, Langston Hughes, Emily Dickinson, T. S. Eliot, other notables. 96pp. 5³⁄₁₆ x 8¼. 0-486-40158-8

101 GREAT SAMURAI PRINTS, Utagawa Kuniyoshi. Kuniyoshi was a master of the warrior woodblock print — and these 18th-century illustrations represent the pinnacle of his craft. Full-color portraits of renowned Japanese samurais pulse with movement, passion, and remarkably fine detail. 112pp. 8⅜ x 11. 0-486-46523-3

ABC OF BALLET, Janet Grosser. Clearly worded, abundantly illustrated little guide defines basic ballet-related terms: arabesque, battement, pas de chat, relevé, sissonne, many others. Pronunciation guide included. Excellent primer. 48pp. 4³⁄₁₆ x 5¾.
0-486-40871-X

ACCESSORIES OF DRESS: An Illustrated Encyclopedia, Katherine Lester and Bess Viola Oerke. Illustrations of hats, veils, wigs, cravats, shawls, shoes, gloves, and other accessories enhance an engaging commentary that reveals the humor and charm of the many-sided story of accessorized apparel. 644 figures and 59 plates. 608pp. 6 ⅛ x 9¼.
0-486-43378-1

ADVENTURES OF HUCKLEBERRY FINN, Mark Twain. Join Huck and Jim as their boyhood adventures along the Mississippi River lead them into a world of excitement, danger, and self-discovery. Humorous narrative, lyrical descriptions of the Mississippi valley, and memorable characters. 224pp. 5³⁄₁₆ x 8¼. 0-486-28061-6

ALICE STARMORE'S BOOK OF FAIR ISLE KNITTING, Alice Starmore. A noted designer from the region of Scotland's Fair Isle explores the history and techniques of this distinctive, stranded-color knitting style and provides copious illustrated instructions for 14 original knitwear designs. 208pp. 8⅜ x 10⅞. 0-486-47218-3

Browse over 9,000 books at www.doverpublications.com

ALICE'S ADVENTURES IN WONDERLAND, Lewis Carroll. Beloved classic about a little girl lost in a topsy-turvy land and her encounters with the White Rabbit, March Hare, Mad Hatter, Cheshire Cat, and other delightfully improbable characters. 42 illustrations by Sir John Tenniel. 96pp. 5¾6 x 8¼. 0-486-27543-4

AMERICA'S LIGHTHOUSES: An Illustrated History, Francis Ross Holland. Profusely illustrated fact-filled survey of American lighthouses since 1716. Over 200 stations — East, Gulf, and West coasts, Great Lakes, Hawaii, Alaska, Puerto Rico, the Virgin Islands, and the Mississippi and St. Lawrence Rivers. 240pp. 8 x 10¾.
0-486-25576-X

AN ENCYCLOPEDIA OF THE VIOLIN, Alberto Bachmann. Translated by Frederick H. Martens. Introduction by Eugene Ysaye. First published in 1925, this renowned reference remains unsurpassed as a source of essential information, from construction and evolution to repertoire and technique. Includes a glossary and 73 illustrations. 496pp. 6⅛ x 9¼. 0-486-46618-3

ANIMALS: 1,419 Copyright-Free Illustrations of Mammals, Birds, Fish, Insects, etc., Selected by Jim Harter. Selected for its visual impact and ease of use, this outstanding collection of wood engravings presents over 1,000 species of animals in extremely lifelike poses. Includes mammals, birds, reptiles, amphibians, fish, insects, and other invertebrates. 284pp. 9 x 12. 0-486-23766-4

THE ANNALS, Tacitus. Translated by Alfred John Church and William Jackson Brodribb. This vital chronicle of Imperial Rome, written by the era's great historian, spans A.D. 14-68 and paints incisive psychological portraits of major figures, from Tiberius to Nero. 416pp. 5¾6 x 8¼. 0-486-45236-0

ANTIGONE, Sophocles. Filled with passionate speeches and sensitive probing of moral and philosophical issues, this powerful and often-performed Greek drama reveals the grim fate that befalls the children of Oedipus. Footnotes. 64pp. 5¾6 x 8 ¼. 0-486-27804-2

ART DECO DECORATIVE PATTERNS IN FULL COLOR, Christian Stoll. Reprinted from a rare 1910 portfolio, 160 sensuous and exotic images depict a breathtaking array of florals, geometrics, and abstracts — all elegant in their stark simplicity. 64pp. 8⅜ x 11. 0-486-44862-2

THE ARTHUR RACKHAM TREASURY: 86 Full-Color Illustrations, Arthur Rackham. Selected and Edited by Jeff A. Menges. A stunning treasury of 86 full-page plates span the famed English artist's career, from *Rip Van Winkle* (1905) to masterworks such as *Undine, A Midsummer Night's Dream*, and *Wind in the Willows* (1939). 96pp. 8⅜ x 11.
0-486-44685-9

THE AUTHENTIC GILBERT & SULLIVAN SONGBOOK, W. S. Gilbert and A. S. Sullivan. The most comprehensive collection available, this songbook includes selections from every one of Gilbert and Sullivan's light operas. Ninety-two numbers are presented uncut and unedited, and in their original keys. 410pp. 9 x 12.
0-486-23482-7

THE AWAKENING, Kate Chopin. First published in 1899, this controversial novel of a New Orleans wife's search for love outside a stifling marriage shocked readers. Today, it remains a first-rate narrative with superb characterization. New introductory Note. 128pp. 5¾6 x 8¼. 0-486-27786-0

BASIC DRAWING, Louis Priscilla. Beginning with perspective, this commonsense manual progresses to the figure in movement, light and shade, anatomy, drapery, composition, trees and landscape, and outdoor sketching. Black-and-white illustrations throughout. 128pp. 8⅜ x 11. 0-486-45815-6

THE BATTLES THAT CHANGED HISTORY, Fletcher Pratt. Historian profiles 16 crucial conflicts, ancient to modern, that changed the course of Western civilization. Gripping accounts of battles led by Alexander the Great, Joan of Arc, Ulysses S. Grant, other commanders. 27 maps. 352pp. 5⅜ x 8½. 0-486-41129-X

BEETHOVEN'S LETTERS, Ludwig van Beethoven. Edited by Dr. A. C. Kalischer. Features 457 letters to fellow musicians, friends, greats, patrons, and literary men. Reveals musical thoughts, quirks of personality, insights, and daily events. Includes 15 plates. 410pp. 5⅜ x 8½. 0-486-22769-3

BERNICE BOBS HER HAIR AND OTHER STORIES, F. Scott Fitzgerald. This brilliant anthology includes 6 of Fitzgerald's most popular stories: "The Diamond as Big as the Ritz," the title tale, "The Offshore Pirate," "The Ice Palace," "The Jelly Bean," and "May Day." 176pp. 5⅜ x 8½. 0-486-47049-0

BESLER'S BOOK OF FLOWERS AND PLANTS: 73 Full-Color Plates from Hortus Eystettensis, 1613, Basilius Besler. Here is a selection of magnificent plates from the *Hortus Eystettensis,* which vividly illustrated and identified the plants, flowers, and trees that thrived in the legendary German garden at Eichstätt. 80pp. 8⅜ x 11. 0-486-46005-3

THE BOOK OF KELLS, Edited by Blanche Cirker. Painstakingly reproduced from a rare facsimile edition, this volume contains full-page decorations, portraits, illustrations, plus a sampling of textual leaves with exquisite calligraphy and ornamentation. 32 full-color illustrations. 32pp. 9⅜ x 12¼. 0-486-24345-1

THE BOOK OF THE CROSSBOW: With an Additional Section on Catapults and Other Siege Engines, Ralph Payne-Gallwey. Fascinating study traces history and use of crossbow as military and sporting weapon, from Middle Ages to modern times. Also covers related weapons: balistas, catapults, Turkish bows, more. Over 240 illustrations. 400pp. 7¼ x 10⅛. 0-486-28720-3

THE BUNGALOW BOOK: Floor Plans and Photos of 112 Houses, 1910, Henry L. Wilson. Here are 112 of the most popular and economic blueprints of the early 20th century — plus an illustration or photograph of each completed house. A wonderful time capsule that still offers a wealth of valuable insights. 160pp. 8⅜ x 11. 0-486-45104-6

THE CALL OF THE WILD, Jack London. A classic novel of adventure, drawn from London's own experiences as a Klondike adventurer, relating the story of a heroic dog caught in the brutal life of the Alaska Gold Rush. Note. 64pp. 5$\frac{3}{16}$ x 8¼. 0-486-26472-6

CANDIDE, Voltaire. Edited by Francois-Marie Arouet. One of the world's great satires since its first publication in 1759. Witty, caustic skewering of romance, science, philosophy, religion, government — nearly all human ideals and institutions. 112pp. 5$\frac{3}{16}$ x 8¼. 0-486-26689-3

CELEBRATED IN THEIR TIME: Photographic Portraits from the George Grantham Bain Collection, Edited by Amy Pastan. With an Introduction by Michael Carlebach. Remarkable portrait gallery features 112 rare images of Albert Einstein, Charlie Chaplin, the Wright Brothers, Henry Ford, and other luminaries from the worlds of politics, art, entertainment, and industry. 128pp. 8⅜ x 11. 0-486-46754-6

CHARIOTS FOR APOLLO: The NASA History of Manned Lunar Spacecraft to 1969, Courtney G. Brooks, James M. Grimwood, and Loyd S. Swenson, Jr. This illustrated history by a trio of experts is the definitive reference on the Apollo spacecraft and lunar modules. It traces the vehicles' design, development, and operation in space. More than 100 photographs and illustrations. 576pp. 6¾ x 9¼. 0-486-46756-2

Browse over 9,000 books at www.doverpublications.com

A CHRISTMAS CAROL, Charles Dickens. This engrossing tale relates Ebenezer Scrooge's ghostly journeys through Christmases past, present, and future and his ultimate transformation from a harsh and grasping old miser to a charitable and compassionate human being. 80pp. 5¾₆ x 8¼. 0-486-26865-9

COMMON SENSE, Thomas Paine. First published in January of 1776, this highly influential landmark document clearly and persuasively argued for American separation from Great Britain and paved the way for the Declaration of Independence. 64pp. 5¾₆ x 8¼. 0-486-29602-4

THE COMPLETE SHORT STORIES OF OSCAR WILDE, Oscar Wilde. Complete texts of "The Happy Prince and Other Tales," "A House of Pomegranates," "Lord Arthur Savile's Crime and Other Stories," "Poems in Prose," and "The Portrait of Mr. W. H." 208pp. 5¾₆ x 8¼. 0-486-45216-6

COMPLETE SONNETS, William Shakespeare. Over 150 exquisite poems deal with love, friendship, the tyranny of time, beauty's evanescence, death, and other themes in language of remarkable power, precision, and beauty. Glossary of archaic terms. 80pp. 5¾₆ x 8¼. 0-486-26686-9

THE COUNT OF MONTE CRISTO: Abridged Edition, Alexandre Dumas. Falsely accused of treason, Edmond Dantès is imprisoned in the bleak Chateau d'If. After a hair-raising escape, he launches an elaborate plot to extract a bitter revenge against those who betrayed him. 448pp. 5¾₆ x 8¼. 0-486-45643-9

CRAFTSMAN BUNGALOWS: Designs from the Pacific Northwest, Yoho & Merritt. This reprint of a rare catalog, showcasing the charming simplicity and cozy style of Craftsman bungalows, is filled with photos of completed homes, plus floor plans and estimated costs. An indispensable resource for architects, historians, and illustrators. 112pp. 10 x 7. 0-486-46875-5

CRAFTSMAN BUNGALOWS: 59 Homes from "The Craftsman," Edited by Gustav Stickley. Best and most attractive designs from Arts and Crafts Movement publication — 1903–1916 — includes sketches, photographs of homes, floor plans, descriptive text. 128pp. 8¼ x 11. 0-486-25829-7

CRIME AND PUNISHMENT, Fyodor Dostoyevsky. Translated by Constance Garnett. Supreme masterpiece tells the story of Raskolnikov, a student tormented by his own thoughts after he murders an old woman. Overwhelmed by guilt and terror, he confesses and goes to prison. 480pp. 5¾₆ x 8¼. 0-486-41587-2

THE DECLARATION OF INDEPENDENCE AND OTHER GREAT DOCUMENTS OF AMERICAN HISTORY: 1775-1865, Edited by John Grafton. Thirteen compelling and influential documents: Henry's "Give Me Liberty or Give Me Death," Declaration of Independence, The Constitution, Washington's First Inaugural Address, The Monroe Doctrine, The Emancipation Proclamation, Gettysburg Address, more. 64pp. 5¾₆ x 8¼. 0-486-41124-9

THE DESERT AND THE SOWN: Travels in Palestine and Syria, Gertrude Bell. "The female Lawrence of Arabia," Gertrude Bell wrote captivating, perceptive accounts of her travels in the Middle East. This intriguing narrative, accompanied by 160 photos, traces her 1905 sojourn in Lebanon, Syria, and Palestine. 368pp. 5⅜ x 8½. 0-486-46876-3

A DOLL'S HOUSE, Henrik Ibsen. Ibsen's best-known play displays his genius for realistic prose drama. An expression of women's rights, the play climaxes when the central character, Nora, rejects a smothering marriage and life in "a doll's house." 80pp. 5¾₆ x 8¼. 0-486-27062-9

DOOMED SHIPS: Great Ocean Liner Disasters, William H. Miller, Jr. Nearly 200 photographs, many from private collections, highlight tales of some of the vessels whose pleasure cruises ended in catastrophe: the *Morro Castle, Normandie, Andrea Doria, Europa,* and many others. 128pp. 8⅛ x 11¾. 0-486-45366-9

THE DORÉ BIBLE ILLUSTRATIONS, Gustave Doré. Detailed plates from the Bible: the Creation scenes, Adam and Eve, horrifying visions of the Flood, the battle sequences with their monumental crowds, depictions of the life of Jesus, 241 plates in all. 241pp. 9 x 12. 0-486-23004-X

DRAWING DRAPERY FROM HEAD TO TOE, Cliff Young. Expert guidance on how to draw shirts, pants, skirts, gloves, hats, and coats on the human figure, including folds in relation to the body, pull and crush, action folds, creases, more. Over 200 drawings. 48pp. 8¼ x 11. 0-486-45591-2

DUBLINERS, James Joyce. A fine and accessible introduction to the work of one of the 20th century's most influential writers, this collection features 15 tales, including a masterpiece of the short-story genre, "The Dead." 160pp. 5³⁄₁₆ x 8¼. 0-486-26870-5

EASY-TO-MAKE POP-UPS, Joan Irvine. Illustrated by Barbara Reid. Dozens of wonderful ideas for three-dimensional paper fun — from holiday greeting cards with moving parts to a pop-up menagerie. Easy-to-follow, illustrated instructions for more than 30 projects. 299 black-and-white illustrations. 96pp. 8⅜ x 11. 0-486-44622-0

EASY-TO-MAKE STORYBOOK DOLLS: A "Novel" Approach to Cloth Dollmaking, Sherralyn St. Clair. Favorite fictional characters come alive in this unique beginner's dollmaking guide. Includes patterns for Pollyanna, Dorothy from *The Wonderful Wizard of Oz*, Mary of *The Secret Garden*, plus easy-to-follow instructions, 263 black-and-white illustrations, and an 8-page color insert. 112pp. 8¼ x 11. 0-486-47360-0

EINSTEIN'S ESSAYS IN SCIENCE, Albert Einstein. Speeches and essays in accessible, everyday language profile influential physicists such as Niels Bohr and Isaac Newton. They also explore areas of physics to which the author made major contributions. 128pp. 5 x 8. 0-486-47011-3

EL DORADO: Further Adventures of the Scarlet Pimpernel, Baroness Orczy. A popular sequel to *The Scarlet Pimpernel*, this suspenseful story recounts the Pimpernel's attempts to rescue the Dauphin from imprisonment during the French Revolution. An irresistible blend of intrigue, period detail, and vibrant characterizations. 352pp. 5³⁄₁₆ x 8¼. 0-486-44026-5

ELEGANT SMALL HOMES OF THE TWENTIES: 99 Designs from a Competition, Chicago Tribune. Nearly 100 designs for five- and six-room houses feature New England and Southern colonials, Normandy cottages, stately Italianate dwellings, and other fascinating snapshots of American domestic architecture of the 1920s. 112pp. 9 x 12. 0-486-46910-7

THE ELEMENTS OF STYLE: The Original Edition, William Strunk, Jr. This is the book that generations of writers have relied upon for timeless advice on grammar, diction, syntax, and other essentials. In concise terms, it identifies the principal requirements of proper style and common errors. 64pp. 5⅜ x 8½. 0-486-44798-7

THE ELUSIVE PIMPERNEL, Baroness Orczy. Robespierre's revolutionaries find their wicked schemes thwarted by the heroic Pimpernel — Sir Percival Blakeney. In this thrilling sequel, Chauvelin devises a plot to eliminate the Pimpernel and his wife. 272pp. 5³⁄₁₆ x 8¼. 0-486-45464-9

AN ENCYCLOPEDIA OF BATTLES: Accounts of Over 1,560 Battles from 1479 B.C. to the Present, David Eggenberger. Essential details of every major battle in recorded history from the first battle of Megiddo in 1479 B.C. to Grenada in 1984. List of battle maps. 99 illustrations. 544pp. 6½ x 9¼.   0-486-24913-1

ENCYCLOPEDIA OF EMBROIDERY STITCHES, INCLUDING CREWEL, Marion Nichols. Precise explanations and instructions, clearly illustrated, on how to work chain, back, cross, knotted, woven stitches, and many more — 178 in all, including Cable Outline, Whipped Satin, and Eyelet Buttonhole. Over 1400 illustrations. 219pp. 8⅜ x 11¼.   0-486-22929-7

ENTER JEEVES: 15 Early Stories, P. G. Wodehouse. Splendid collection contains first 8 stories featuring Bertie Wooster, the deliciously dim aristocrat and Jeeves, his brainy, imperturbable manservant. Also, the complete Reggie Pepper (Bertie's prototype) series. 288pp. 5⅜ x 8½.   0-486-29717-9

ERIC SLOANE'S AMERICA: Paintings in Oil, Michael Wigley. With a Foreword by Mimi Sloane. Eric Sloane's evocative oils of America's landscape and material culture shimmer with immense historical and nostalgic appeal. This original hardcover collection gathers nearly a hundred of his finest paintings, with subjects ranging from New England to the American Southwest. 128pp. 10⅞ x 9.
0-486-46525-X

ETHAN FROME, Edith Wharton. Classic story of wasted lives, set against a bleak New England background. Superbly delineated characters in a hauntingly grim tale of thwarted love. Considered by many to be Wharton's masterpiece. 96pp. 5³⁄₁₆ x 8¼.
0-486-26690-7

THE EVERLASTING MAN, G. K. Chesterton. Chesterton's view of Christianity — as a blend of philosophy and mythology, satisfying intellect and spirit — applies to his brilliant book, which appeals to readers' heads as well as their hearts. 288pp. 5⅜ x 8½.
0-486-46036-3

THE FIELD AND FOREST HANDY BOOK, Daniel Beard. Written by a co-founder of the Boy Scouts, this appealing guide offers illustrated instructions for building kites, birdhouses, boats, igloos, and other fun projects, plus numerous helpful tips for campers. 448pp. 5³⁄₁₆ x 8¼.   0-486-46191-2

FINDING YOUR WAY WITHOUT MAP OR COMPASS, Harold Gatty. Useful, instructive manual shows would-be explorers, hikers, bikers, scouts, sailors, and survivalists how to find their way outdoors by observing animals, weather patterns, shifting sands, and other elements of nature. 288pp. 5⅜ x 8½.   0-486-40613-X

FIRST FRENCH READER: A Beginner's Dual-Language Book, Edited and Translated by Stanley Appelbaum. This anthology introduces 50 legendary writers — Voltaire, Balzac, Baudelaire, Proust, more — through passages from *The Red and the Black*, *Les Misérables, Madame Bovary*, and other classics. Original French text plus English translation on facing pages. 240pp. 5⅜ x 8½.   0-486-46178-5

FIRST GERMAN READER: A Beginner's Dual-Language Book, Edited by Harry Steinhauer. Specially chosen for their power to evoke German life and culture, these short, simple readings include poems, stories, essays, and anecdotes by Goethe, Hesse, Heine, Schiller, and others. 224pp. 5⅜ x 8½.   0-486-46179-3

FIRST SPANISH READER: A Beginner's Dual-Language Book, Angel Flores. Delightful stories, other material based on works of Don Juan Manuel, Luis Taboada, Ricardo Palma, other noted writers. Complete faithful English translations on facing pages. Exercises. 176pp. 5⅜ x 8½.   0-486-25810-6

Browse over 9,000 books at www.doverpublications.com

FIVE ACRES AND INDEPENDENCE, Maurice G. Kains. Great back-to-the-land classic explains basics of self-sufficient farming. The one book to get. 95 illustrations. 397pp. 5⅜ x 8½.
0-486-20974-1

FLAGG'S SMALL HOUSES: Their Economic Design and Construction, 1922, Ernest Flagg. Although most famous for his skyscrapers, Flagg was also a proponent of the well-designed single-family dwelling. His classic treatise features innovations that save space, materials, and cost. 526 illustrations. 160pp. 9⅜ x 12¼.
0-486-45197-6

FLATLAND: A Romance of Many Dimensions, Edwin A. Abbott. Classic of science (and mathematical) fiction — charmingly illustrated by the author — describes the adventures of A. Square, a resident of Flatland, in Spaceland (three dimensions), Lineland (one dimension), and Pointland (no dimensions). 96pp. 5³⁄₁₆ x 8¼.
0-486-27263-X

FRANKENSTEIN, Mary Shelley. The story of Victor Frankenstein's monstrous creation and the havoc it caused has enthralled generations of readers and inspired countless writers of horror and suspense. With the author's own 1831 introduction. 176pp. 5³⁄₁₆ x 8¼.
0-486-28211-2

THE GARGOYLE BOOK: 572 Examples from Gothic Architecture, Lester Burbank Bridaham. Dispelling the conventional wisdom that French Gothic architectural flourishes were born of despair or gloom, Bridaham reveals the whimsical nature of these creations and the ingenious artisans who made them. 572 illustrations. 224pp. 8⅜ x 11.
0-486-44754-5

THE GIFT OF THE MAGI AND OTHER SHORT STORIES, O. Henry. Sixteen captivating stories by one of America's most popular storytellers. Included are such classics as "The Gift of the Magi," "The Last Leaf," and "The Ransom of Red Chief." Publisher's Note. 96pp. 5³⁄₁₆ x 8¼.
0-486-27061-0

THE GOETHE TREASURY: Selected Prose and Poetry, Johann Wolfgang von Goethe. Edited, Selected, and with an Introduction by Thomas Mann. In addition to his lyric poetry, Goethe wrote travel sketches, autobiographical studies, essays, letters, and proverbs in rhyme and prose. This collection presents outstanding examples from each genre. 368pp. 5⅜ x 8½.
0-486-44780-4

GREAT EXPECTATIONS, Charles Dickens. Orphaned Pip is apprenticed to the dirty work of the forge but dreams of becoming a gentleman — and one day finds himself in possession of "great expectations." Dickens' finest novel. 400pp. 5³⁄₁₆ x 8¼.
0-486-41586-4

GREAT WRITERS ON THE ART OF FICTION: From Mark Twain to Joyce Carol Oates, Edited by James Daley. An indispensable source of advice and inspiration, this anthology features essays by Henry James, Kate Chopin, Willa Cather, Sinclair Lewis, Jack London, Raymond Chandler, Raymond Carver, Eudora Welty, and Kurt Vonnegut, Jr. 192pp. 5⅜ x 8½.
0-486-45128-3

HAMLET, William Shakespeare. The quintessential Shakespearean tragedy, whose highly charged confrontations and anguished soliloquies probe depths of human feeling rarely sounded in any art. Reprinted from an authoritative British edition complete with illuminating footnotes. 128pp. 5³⁄₁₆ x 8¼.
0-486-27278-8

THE HAUNTED HOUSE, Charles Dickens. A Yuletide gathering in an eerie country retreat provides the backdrop for Dickens and his friends — including Elizabeth Gaskell and Wilkie Collins — who take turns spinning supernatural yarns. 144pp. 5⅜ x 8½.
0-486-46309-5

HEART OF DARKNESS, Joseph Conrad. Dark allegory of a journey up the Congo River and the narrator's encounter with the mysterious Mr. Kurtz. Masterly blend of adventure, character study, psychological penetration. For many, Conrad's finest, most enigmatic story. 80pp. 5³⁄₁₆ x 8¼. 0-486-26464-5

HENSON AT THE NORTH POLE, Matthew A. Henson. This thrilling memoir by the heroic African-American who was Peary's companion through two decades of Arctic exploration recounts a tale of danger, courage, and determination. "Fascinating and exciting." — *Commonweal.* 128pp. 5⅜ x 8½. 0-486-45472-X

HISTORIC COSTUMES AND HOW TO MAKE THEM, Mary Fernald and E. Shenton. Practical, informative guidebook shows how to create everything from short tunics worn by Saxon men in the fifth century to a lady's bustle dress of the late 1800s. 81 illustrations. 176pp. 5⅜ x 8½. 0-486-44906-8

THE HOUND OF THE BASKERVILLES, Arthur Conan Doyle. A deadly curse in the form of a legendary ferocious beast continues to claim its victims from the Baskerville family until Holmes and Watson intervene. Often called the best detective story ever written. 128pp. 5³⁄₁₆ x 8¼. 0-486-28214-7

THE HOUSE BEHIND THE CEDARS, Charles W. Chesnutt. Originally published in 1900, this groundbreaking novel by a distinguished African-American author recounts the drama of a brother and sister who "pass for white" during the dangerous days of Reconstruction. 208pp. 5⅜ x 8½. 0-486-46144-0

THE HUMAN FIGURE IN MOTION, Eadweard Muybridge. The 4,789 photographs in this definitive selection show the human figure — models almost all undraped — engaged in over 160 different types of action: running, climbing stairs, etc. 390pp. 7⅞ x 10⅝. 0-486-20204-6

THE IMPORTANCE OF BEING EARNEST, Oscar Wilde. Wilde's witty and buoyant comedy of manners, filled with some of literature's most famous epigrams, reprinted from an authoritative British edition. Considered Wilde's most perfect work. 64pp. 5³⁄₁₆ x 8¼. 0-486-26478-5

THE INFERNO, Dante Alighieri. Translated and with notes by Henry Wadsworth Longfellow. The first stop on Dante's famous journey from Hell to Purgatory to Paradise, this 14th-century allegorical poem blends vivid and shocking imagery with graceful lyricism. Translated by the beloved 19th-century poet, Henry Wadsworth Longfellow. 256pp. 5³⁄₁₆ x 8¼. 0-486-44288-8

JANE EYRE, Charlotte Brontë. Written in 1847, *Jane Eyre* tells the tale of an orphan girl's progress from the custody of cruel relatives to an oppressive boarding school and its culmination in a troubled career as a governess. 448pp. 5³⁄₁₆ x 8¼.
0-486-42449-9

JAPANESE WOODBLOCK FLOWER PRINTS, Tanigami Kônan. Extraordinary collection of Japanese woodblock prints by a well-known artist features 120 plates in brilliant color. Realistic images from a rare edition include daffodils, tulips, and other familiar and unusual flowers. 128pp. 11 x 8¼. 0-486-46442-3

JEWELRY MAKING AND DESIGN, Augustus F. Rose and Antonio Cirino. Professional secrets of jewelry making are revealed in a thorough, practical guide. Over 200 illustrations. 306pp. 5⅜ x 8½. 0-486-21750-7

JULIUS CAESAR, William Shakespeare. Great tragedy based on Plutarch's account of the lives of Brutus, Julius Caesar and Mark Antony. Evil plotting, ringing oratory, high tragedy with Shakespeare's incomparable insight, dramatic power. Explanatory footnotes. 96pp. 5³⁄₁₆ x 8¼. 0-486-26876-4

Browse over 9,000 books at www.doverpublications.com

THE JUNGLE, Upton Sinclair. 1906 bestseller shockingly reveals intolerable labor practices and working conditions in the Chicago stockyards as it tells the grim story of a Slavic family that emigrates to America full of optimism but soon faces despair. 320pp. 5³⁄₁₆ x 8¼. 0-486-41923-1

THE KINGDOM OF GOD IS WITHIN YOU, Leo Tolstoy. The soul-searching book that inspired Gandhi to embrace the concept of passive resistance, Tolstoy's 1894 polemic clearly outlines a radical, well-reasoned revision of traditional Christian thinking. 352pp. 5³⁄₁₆ x 8¼. 0-486-45138-0

THE LADY OR THE TIGER?: and Other Logic Puzzles, Raymond M. Smullyan. Created by a renowned puzzle master, these whimsically themed challenges involve paradoxes about probability, time, and change; metapuzzles; and self-referentiality. Nineteen chapters advance in difficulty from relatively simple to highly complex. 1982 edition. 240pp. 5⅜ x 8½. 0-486-47027-X

LEAVES OF GRASS: The Original 1855 Edition, Walt Whitman. Whitman's immortal collection includes some of the greatest poems of modern times, including his masterpiece, "Song of Myself." Shattering standard conventions, it stands as an unabashed celebration of body and nature. 128pp. 5³⁄₁₆ x 8¼. 0-486-45676-5

LES MISÉRABLES, Victor Hugo. Translated by Charles E. Wilbour. Abridged by James K. Robinson. A convict's heroic struggle for justice and redemption plays out against a fiery backdrop of the Napoleonic wars. This edition features the excellent original translation and a sensitive abridgment. 304pp. 6⅛ x 9¼. 0-486-45789-3

LILITH: A Romance, George MacDonald. In this novel by the father of fantasy literature, a man travels through time to meet Adam and Eve and to explore humanity's fall from grace and ultimate redemption. 240pp. 5⅜ x 8½. 0-486-46818-6

THE LOST LANGUAGE OF SYMBOLISM, Harold Bayley. This remarkable book reveals the hidden meaning behind familiar images and words, from the origins of Santa Claus to the fleur-de-lys, drawing from mythology, folklore, religious texts, and fairy tales. 1,418 illustrations. 784pp. 5⅜ x 8½. 0-486-44787-1

MACBETH, William Shakespeare. A Scottish nobleman murders the king in order to succeed to the throne. Tortured by his conscience and fearful of discovery, he becomes tangled in a web of treachery and deceit that ultimately spells his doom. 96pp. 5³⁄₁₆ x 8¼. 0-486-27802-6

MAKING AUTHENTIC CRAFTSMAN FURNITURE: Instructions and Plans for 62 Projects, Gustav Stickley. Make authentic reproductions of handsome, functional, durable furniture: tables, chairs, wall cabinets, desks, a hall tree, and more. Construction plans with drawings, schematics, dimensions, and lumber specs reprinted from 1900s The Craftsman magazine. 128pp. 8⅛ x 11. 0-486-25000-8

MATHEMATICS FOR THE NONMATHEMATICIAN, Morris Kline. Erudite and entertaining overview follows development of mathematics from ancient Greeks to present. Topics include logic and mathematics, the fundamental concept, differential calculus, probability theory, much more. Exercises and problems. 641pp. 5⅜ x 8½. 0-486-24823-2

MEMOIRS OF AN ARABIAN PRINCESS FROM ZANZIBAR, Emily Ruete. This 19th-century autobiography offers a rare inside look at the society surrounding a sultan's palace. A real-life princess in exile recalls her vanished world of harems, slave trading, and court intrigues. 288pp. 5⅜ x 8½. 0-486-47121-7

Browse over 9,000 books at www.doverpublications.com

THE METAMORPHOSIS AND OTHER STORIES, Franz Kafka. Excellent new English translations of title story (considered by many critics Kafka's most perfect work), plus "The Judgment," "In the Penal Colony," "A Country Doctor," and "A Report to an Academy." Note. 96pp. 5³⁄₁₆ x 8¼.   0-486-29030-1

MICROSCOPIC ART FORMS FROM THE PLANT WORLD, R. Anheisser. From undulating curves to complex geometrics, a world of fascinating images abound in this classic, illustrated survey of microscopic plants. Features 400 detailed illustrations of nature's minute but magnificent handiwork. The accompanying CD-ROM includes all of the images in the book. 128pp. 9 x 9.   0-486-46013-4

A MIDSUMMER NIGHT'S DREAM, William Shakespeare. Among the most popular of Shakespeare's comedies, this enchanting play humorously celebrates the vagaries of love as it focuses upon the intertwined romances of several pairs of lovers. Explanatory footnotes. 80pp. 5³⁄₁₆ x 8¼.   0-486-27067-X

THE MONEY CHANGERS, Upton Sinclair. Originally published in 1908, this cautionary novel from the author of *The Jungle* explores corruption within the American system as a group of power brokers joins forces for personal gain, triggering a crash on Wall Street. 192pp. 5⅜ x 8½.   0-486-46917-4

THE MOST POPULAR HOMES OF THE TWENTIES, William A. Radford. With a New Introduction by Daniel D. Reiff. Based on a rare 1925 catalog, this architectural showcase features floor plans, construction details, and photos of 26 homes, plus articles on entrances, porches, garages, and more. 250 illustrations, 21 color plates. 176pp. 8⅜ x 11.   0-486-47028-8

MY 66 YEARS IN THE BIG LEAGUES, Connie Mack. With a New Introduction by Rich Westcott. A Founding Father of modern baseball, Mack holds the record for most wins — and losses — by a major league manager. Enhanced by 70 photographs, his warmhearted autobiography is populated by many legends of the game. 288pp. 5⅜ x 8½.   0-486-47184-5

NARRATIVE OF THE LIFE OF FREDERICK DOUGLASS, Frederick Douglass. Douglass's graphic depictions of slavery, harrowing escape to freedom, and life as a newspaper editor, eloquent orator, and impassioned abolitionist. 96pp. 5³⁄₁₆ x 8¼.   0-486-28499-9

THE NIGHTLESS CITY: Geisha and Courtesan Life in Old Tokyo, J. E. de Becker. This unsurpassed study from 100 years ago ventured into Tokyo's red-light district to survey geisha and courtesan life and offer meticulous descriptions of training, dress, social hierarchy, and erotic practices. 49 black-and-white illustrations; 2 maps. 496pp. 5⅜ x 8½.   0-486-45563-7

THE ODYSSEY, Homer. Excellent prose translation of ancient epic recounts adventures of the homeward-bound Odysseus. Fantastic cast of gods, giants, cannibals, sirens, other supernatural creatures — true classic of Western literature. 256pp. 5³⁄₁₆ x 8¼.   0-486-40654-7

OEDIPUS REX, Sophocles. Landmark of Western drama concerns the catastrophe that ensues when King Oedipus discovers he has inadvertently killed his father and married his mother. Masterly construction, dramatic irony. Explanatory footnotes. 64pp. 5³⁄₁₆ x 8¼.   0-486-26877-2

ONCE UPON A TIME: The Way America Was, Eric Sloane. Nostalgic text and drawings brim with gentle philosophies and descriptions of how we used to live — self-sufficiently — on the land, in homes, and among the things built by hand. 44 line illustrations. 64pp. 8⅜ x 11.   0-486-44411-2

Browse over 9,000 books at www.doverpublications.com

ONE OF OURS, Willa Cather. The Pulitzer Prize–winning novel about a young Nebraskan looking for something to believe in. Alienated from his parents, rejected by his wife, he finds his destiny on the bloody battlefields of World War I. 352pp. 5³⁄₁₆ x 8¼. 0-486-45599-8

ORIGAMI YOU CAN USE: 27 Practical Projects, Rick Beech. Origami models can be more than decorative, and this unique volume shows how! The 27 practical projects include a CD case, frame, napkin ring, and dish. Easy instructions feature 400 two-color illustrations. 96pp. 8¼ x 11. 0-486-47057-1

OTHELLO, William Shakespeare. Towering tragedy tells the story of a Moorish general who earns the enmity of his ensign Iago when he passes him over for a promotion. Masterly portrait of an archvillain. Explanatory footnotes. 112pp. 5³⁄₁₆ x 8¼.
0-486-29097-2

PARADISE LOST, John Milton. Notes by John A. Himes. First published in 1667, *Paradise Lost* ranks among the greatest of English literature's epic poems. It's a sublime retelling of Adam and Eve's fall from grace and expulsion from Eden. Notes by John A. Himes. 480pp. 5³⁄₁₆ x 8¼. 0-486-44287-X

PASSING, Nella Larsen. Married to a successful physician and prominently ensconced in society, Irene Redfield leads a charmed existence — until a chance encounter with a childhood friend who has been "passing for white." 112pp. 5⅜ x 8½. 0-486-43713-2

PERSPECTIVE DRAWING FOR BEGINNERS, Len A. Doust. Doust carefully explains the roles of lines, boxes, and circles, and shows how visualizing shapes and forms can be used in accurate depictions of perspective. One of the most concise introductions available. 33 illustrations. 64pp. 5⅜ x 8½. 0-486-45149-6

PERSPECTIVE MADE EASY, Ernest R. Norling. Perspective is easy; yet, surprisingly few artists know the simple rules that make it so. Remedy that situation with this simple, step-by-step book, the first devoted entirely to the topic. 256 illustrations. 224pp. 5⅜ x 8½. 0-486-40473-0

THE PICTURE OF DORIAN GRAY, Oscar Wilde. Celebrated novel involves a handsome young Londoner who sinks into a life of depravity. His body retains perfect youth and vigor while his recent portrait reflects the ravages of his crime and sensuality. 176pp. 5³⁄₁₆ x 8¼. 0-486-27807-7

PRIDE AND PREJUDICE, Jane Austen. One of the most universally loved and admired English novels, an effervescent tale of rural romance transformed by Jane Austen's art into a witty, shrewdly observed satire of English country life. 272pp. 5³⁄₁₆ x 8¼.
0-486-28473-5

THE PRINCE, Niccolò Machiavelli. Classic, Renaissance-era guide to acquiring and maintaining political power. Today, nearly 500 years after it was written, this calculating prescription for autocratic rule continues to be much read and studied. 80pp. 5³⁄₁₆ x 8¼. 0-486-27274-5

QUICK SKETCHING, Carl Cheek. A perfect introduction to the technique of "quick sketching." Drawing upon an artist's immediate emotional responses, this is an extremely effective means of capturing the essential form and features of a subject. More than 100 black-and-white illustrations throughout. 48pp. 11 x 8¼.
0-486-46608-6

RANCH LIFE AND THE HUNTING TRAIL, Theodore Roosevelt. Illustrated by Frederic Remington. Beautifully illustrated by Remington, Roosevelt's celebration of the Old West recounts his adventures in the Dakota Badlands of the 1880s, from round-ups to Indian encounters to hunting bighorn sheep. 208pp. 6¼ x 9¼. 0-486-47340-6

THE RED BADGE OF COURAGE, Stephen Crane. Amid the nightmarish chaos of a Civil War battle, a young soldier discovers courage, humility, and, perhaps, wisdom. Uncanny re-creation of actual combat. Enduring landmark of American fiction. 112pp. 5³⁄₁₆ x 8¼. 0-486-26465-3

RELATIVITY SIMPLY EXPLAINED, Martin Gardner. One of the subject's clearest, most entertaining introductions offers lucid explanations of special and general theories of relativity, gravity, and spacetime, models of the universe, and more. 100 illustrations. 224pp. 5⅜ x 8½. 0-486-29315-7

REMBRANDT DRAWINGS: 116 Masterpieces in Original Color, Rembrandt van Rijn. This deluxe hardcover edition features drawings from throughout the Dutch master's prolific career. Informative captions accompany these beautifully reproduced landscapes, biblical vignettes, figure studies, animal sketches, and portraits. 128pp. 8⅜ x 11. 0-486-46149-1

THE ROAD NOT TAKEN AND OTHER POEMS, Robert Frost. A treasury of Frost's most expressive verse. In addition to the title poem: "An Old Man's Winter Night," "In the Home Stretch," "Meeting and Passing," "Putting in the Seed," many more. All complete and unabridged. 64pp. 5³⁄₁₆ x 8¼. 0-486-27550-7

ROMEO AND JULIET, William Shakespeare. Tragic tale of star-crossed lovers, feuding families and timeless passion contains some of Shakespeare's most beautiful and lyrical love poetry. Complete, unabridged text with explanatory footnotes. 96pp. 5³⁄₁₆ x 8¼. 0-486-27557-4

SANDITON AND THE WATSONS: Austen's Unfinished Novels, Jane Austen. Two tantalizing incomplete stories revisit Austen's customary milieu of courtship and venture into new territory, amid guests at a seaside resort. Both are worth reading for pleasure and study. 112pp. 5⅜ x 8½. 0-486-45793-1

THE SCARLET LETTER, Nathaniel Hawthorne. With stark power and emotional depth, Hawthorne's masterpiece explores sin, guilt, and redemption in a story of adultery in the early days of the Massachusetts Colony. 192pp. 5³⁄₁₆ x 8¼.
0-486-28048-9

THE SEASONS OF AMERICA PAST, Eric Sloane. Seventy-five illustrations depict cider mills and presses, sleds, pumps, stump-pulling equipment, plows, and other elements of America's rural heritage. A section of old recipes and household hints adds additional color. 160pp. 8⅜ x 11. 0-486-44220-9

SELECTED CANTERBURY TALES, Geoffrey Chaucer. Delightful collection includes the General Prologue plus three of the most popular tales: "The Knight's Tale," "The Miller's Prologue and Tale," and "The Wife of Bath's Prologue and Tale." In modern English. 144pp. 5³⁄₁₆ x 8¼. 0-486-28241-4

SELECTED POEMS, Emily Dickinson. Over 100 best-known, best-loved poems by one of America's foremost poets, reprinted from authoritative early editions. No comparable edition at this price. Index of first lines. 64pp. 5³⁄₁₆ x 8¼. 0-486-26466-1

SIDDHARTHA, Hermann Hesse. Classic novel that has inspired generations of seekers. Blending Eastern mysticism and psychoanalysis, Hesse presents a strikingly original view of man and culture and the arduous process of self-discovery, reconciliation, harmony, and peace. 112pp. 5³⁄₁₆ x 8¼. 0-486-40653-9

SKETCHING OUTDOORS, Leonard Richmond. This guide offers beginners step-by-step demonstrations of how to depict clouds, trees, buildings, and other outdoor sights. Explanations of a variety of techniques include shading and constructional drawing. 48pp. 11 x 8¼. 0-486-46922-0

CATALOG OF DOVER BOOKS

SMALL HOUSES OF THE FORTIES: With Illustrations and Floor Plans, Harold E. Group. 56 floor plans and elevations of houses that originally cost less than $15,000 to build. Recommended by financial institutions of the era, they range from Colonials to Cape Cods. 144pp. 8⅜ x 11.　0-486-45598-X

SOME CHINESE GHOSTS, Lafcadio Hearn. Rooted in ancient Chinese legends, these richly atmospheric supernatural tales are recounted by an expert in Oriental lore. Their originality, power, and literary charm will captivate readers of all ages. 96pp. 5⅜ x 8½.　0-486-46306-0

SONGS FOR THE OPEN ROAD: Poems of Travel and Adventure, Edited by The American Poetry & Literacy Project. More than 80 poems by 50 American and British masters celebrate real and metaphorical journeys. Poems by Whitman, Byron, Millay, Sandburg, Langston Hughes, Emily Dickinson, Robert Frost, Shelley, Tennyson, Yeats, many others. Note. 80pp. 5³⁄₁₆ x 8¼.　0-486-40646-6

SPOON RIVER ANTHOLOGY, Edgar Lee Masters. An American poetry classic, in which former citizens of a mythical midwestern town speak touchingly from the grave of the thwarted hopes and dreams of their lives. 144pp. 5³⁄₁₆ x 8¼.
0-486-27275-3

STAR LORE: Myths, Legends, and Facts, William Tyler Olcott. Captivating retellings of the origins and histories of ancient star groups include Pegasus, Ursa Major, Pleiades, signs of the zodiac, and other constellations. "Classic." — *Sky & Telescope.* 58 illustrations. 544pp. 5⅜ x 8½.　0-486-43581-4

THE STRANGE CASE OF DR. JEKYLL AND MR. HYDE, Robert Louis Stevenson. This intriguing novel, both fantasy thriller and moral allegory, depicts the struggle of two opposing personalities — one essentially good, the other evil — for the soul of one man. 64pp. 5³⁄₁₆ x 8¼.　0-486-26688-5

SURVIVAL HANDBOOK: The Official U.S. Army Guide, Department of the Army. This special edition of the Army field manual is geared toward civilians. An essential companion for campers and all lovers of the outdoors, it constitutes the most authoritative wilderness guide. 288pp. 5³⁄₁₆ x 8¼.　0-486-46184-X

A TALE OF TWO CITIES, Charles Dickens. Against the backdrop of the French Revolution, Dickens unfolds his masterpiece of drama, adventure, and romance about a man falsely accused of treason. Excitement and derring-do in the shadow of the guillotine. 304pp. 5³⁄₁₆ x 8¼.　0-486-40651-2

TEN PLAYS, Anton Chekhov. *The Sea Gull, Uncle Vanya, The Three Sisters, The Cherry Orchard,* and *Ivanov,* plus 5 one-act comedies: *The Anniversary, An Unwilling Martyr, The Wedding, The Bear,* and *The Proposal.* 336pp. 5³⁄₁₆ x 8¼.　0-486-46560-8

THE FLYING INN, G. K. Chesterton. Hilarious romp in which pub owner Humphrey Hump and friend take to the road in a donkey cart filled with rum and cheese, inveighing against Prohibition and other "oppressive forms of modernity." 320pp. 5⅜ x 8½.　0-486-41910-X

THIRTY YEARS THAT SHOOK PHYSICS: The Story of Quantum Theory, George Gamow. Lucid, accessible introduction to the influential theory of energy and matter features careful explanations of Dirac's anti-particles, Bohr's model of the atom, and much more. Numerous drawings. 1966 edition. 240pp. 5⅜ x 8½. 0-486-24895-X

TREASURE ISLAND, Robert Louis Stevenson. Classic adventure story of a perilous sea journey, a mutiny led by the infamous Long John Silver, and a lethal scramble for buried treasure — seen through the eyes of cabin boy Jim Hawkins. 160pp. 5³⁄₁₆ x 8¼.
0-486-27559-0

Browse over 9,000 books at www.doverpublications.com

THE TRIAL, Franz Kafka. Translated by David Wyllie. From its gripping first sentence onward, this novel exemplifies the term "Kafkaesque." Its darkly humorous narrative recounts a bank clerk's entrapment in a bureaucratic maze, based on an undisclosed charge. 176pp. 5³⁄₁₆ x 8¼. 0-486-47061-X

THE TURN OF THE SCREW, Henry James. Gripping ghost story by great novelist depicts the sinister transformation of 2 innocent children into flagrant liars and hypocrites. An elegantly told tale of unspoken horror and psychological terror. 96pp. 5³⁄₁₆ x 8¼. 0-486-26684-2

UP FROM SLAVERY, Booker T. Washington. Washington (1856-1915) rose to become the most influential spokesman for African-Americans of his day. In this eloquently written book, he describes events in a remarkable life that began in bondage and culminated in worldwide recognition. 160pp. 5³⁄₁₆ x 8¼. 0-486-28738-6

VICTORIAN HOUSE DESIGNS IN AUTHENTIC FULL COLOR: 75 Plates from the "Scientific American – Architects and Builders Edition," 1885-1894, Edited by Blanche Cirker. Exquisitely detailed, exceptionally handsome designs for an enormous variety of attractive city dwellings, spacious suburban and country homes, charming "cottages" and other structures — all accompanied by perspective views and floor plans. 80pp. 9¼ x 12¼. 0-486-29438-2

VILLETTE, Charlotte Brontë. Acclaimed by Virginia Woolf as "Brontë's finest novel," this moving psychological study features a remarkably modern heroine who abandons her native England for a new life as a schoolteacher in Belgium. 480pp. 5³⁄₁₆ x 8¼. 0-486-45557-2

THE VOYAGE OUT, Virginia Woolf. A moving depiction of the thrills and confusion of youth, Woolf's acclaimed first novel traces a shipboard journey to South America for a captivating exploration of a woman's growing self-awareness. 288pp. 5³⁄₁₆ x 8¼. 0-486-45005-8

WALDEN; OR, LIFE IN THE WOODS, Henry David Thoreau. Accounts of Thoreau's daily life on the shores of Walden Pond outside Concord, Massachusetts, are interwoven with musings on the virtues of self-reliance and individual freedom, on society, government, and other topics. 224pp. 5³⁄₁₆ x 8¼. 0-486-28495-6

WILD PILGRIMAGE: A Novel in Woodcuts, Lynd Ward. Through startling engravings shaded in black and red, Ward wordlessly tells the story of a man trapped in an industrial world, struggling between the grim reality around him and the fantasies his imagination creates. 112pp. 6⅛ x 9¼. 0-486-46583-7

WILLY POGÁNY REDISCOVERED, Willy Pogány. Selected and Edited by Jeff A. Menges. More than 100 color and black-and-white Art Nouveau–style illustrations from fairy tales and adventure stories include scenes from Wagner's "Ring" cycle, *The Rime of the Ancient Mariner, Gulliver's Travels,* and *Faust.* 144pp. 8⅜ x 11. 0-486-47046-6

WOOLLY THOUGHTS: Unlock Your Creative Genius with Modular Knitting, Pat Ashforth and Steve Plummer. Here's the revolutionary way to knit — easy, fun, and foolproof! Beginners and experienced knitters need only master a single stitch to create their own designs with patchwork squares. More than 100 illustrations. 128pp. 6½ x 9¼. 0-486-46084-3

WUTHERING HEIGHTS, Emily Brontë. Somber tale of consuming passions and vengeance — played out amid the lonely English moors — recounts the turbulent and tempestuous love story of Cathy and Heathcliff. Poignant and compelling. 256pp. 5³⁄₁₆ x 8¼. 0-486-29256-8